Dadao Lu
Jie Fan

Regional Development Research in China: A Roadmap to 2050

Dadao Lu
Jie Fan
Editors

Regional Development Research in China: A Roadmap to 2050

With 21 figures

Editors

Dadao Lu
Institute of Geographic Sciences and
Natural Resources Research, CAS
100101, Beijing, China
E-mail: ludd@igsnrr.ac.cn

Jie Fan
Institute of Geographic Sciences and
Natural Resources Research, CAS
100101, Beijing, China
E-mail: fanj@igsnrr.ac.cn

ISBN 978-7-03-027509-7
Science Press Beijing

ISBN 978-3-642-13994-9 e-ISBN 978-3-642-13995-6
Springer Heidelberg Dordrecht London New York

Library of Congress Control Number: 2010929190

Cover design: Frido Steinen-Broo, EStudio Calamar, Spain

Printed on acid-free paper

Springer is a part of Springer Science+Business Media (www.springer.com)

Research Group on Regional Development Research of the Chinese Academy of Sciences

Group leaders: Dadao Lu Institute of Geographic Sciences and Natural Resources Research, CAS

Jie Fan Institute of Geographic Sciences and Natural Resources Research, CAS

Members: (In the alphabetical order of Chinese surname)

Dong Chen	Institute of Geographic Sciences and Natural Resources Research, CAS
Guojie Chen	Chengdu Institute of Mountain Hazards and Environment, CAS
Tian Chen	Institute of Geographic Sciences and Natural Resources Research, CAS
Wen Chen	Nanjing Institute of Geography and Limnology, CAS
Wei Deng	Chengdu Institute of Mountain Hazards and Environment, CAS
Suocheng Dong	Institute of Geographic Sciences and Natural Resources Research, CAS
Xuejun Duan	Nanjing Institute of Geography and Limnology, CAS
Chuanglin Fang	Institute of Geographic Sciences and Natural Resources Research, CAS
Yiping Fang	Chengdu Institute of Mountain Hazards and Environment, CAS
Xiaolu Gao	Institute of Geographic Sciences and Natural Resources Research, CAS
Zhiyong Hu	Institute of Geographic Sciences and Natural Resources Research, CAS
Jinchuan Huang	Institute of Geographic Sciences and Natural Resources Research, CAS
Fengjun Jin	Institute of Geographic Sciences and Natural Resources Research, CAS
Yu Li	Institute of Geographic Sciences and Natural Resources Research, CAS
Jiaming Liu	Institute of Geographic Sciences and Natural Resources Research, CAS
Weidong Liu	Institute of Geographic Sciences and Natural Resources Research, CAS
Yansui Liu	Institute of Geographic Sciences and Natural Resources Research, CAS
Hualou Long	Institute of Geographic Sciences and Natural Resources Research, CAS
Li Ma	Institute of Geographic Sciences and Natural Resources Research, CAS
Yanji Ma	Northeast Institute of Geography and Agroecology, CAS
Hanying Mao	Institute of Geographic Sciences and Natural Resources Research, CAS
Zhixiang She	Nanjing Branch, CAS
Wei Sun	Institute of Geographic Sciences and Natural Resources Research, CAS
Anjun Tao	Institute of Geographic Sciences and Natural Resources Research, CAS
Chengjin Wang	Institute of Geographic Sciences and Natural Resources Research, CAS
Zheng Wang	Institute of Policy and Management, CAS
Guishan Yang	Nanjing Institute of Geography and Limnology, CAS

Zhaoping Yang	Xinjiang Institute of Ecology and Geography, CAS
Xiaogan Yu	Nanjing Institute of Geography and Limnology, CAS
Pingyu Zhang	Northeast Institute of Geography and Agroecology Ecology, CAS
Wenzhong Zhang	Institute of Geographic Sciences and Natural Resources Research, CAS
Xiaolei Zhang	Xinjiang Institute of Ecology and Geography, CAS
Linsheng Zhong	Institute of Geographic Sciences and Natural Resources Research, CAS

English Language Reviewers:

Zhiyong Hu	Institute of Geographic Sciences and Natural Resources Research, CAS
Zhigao Liu	Institute of Geographic Sciences and Natural Resources Research, CAS
Qing Ren	Institute of Geographic Sciences and Natural Resources Research, CAS

Foreword to the Roadmaps 2050*

China's modernization is viewed as a transformative revolution in the human history of modernization. As such, the Chinese Academy of Sciences (CAS) decided to give higher priority to the research on the science and technology (S&T) roadmap for priority areas in China's modernization process. What is the purpose? And why is it? Is it a must? I think those are substantial and significant questions to start things forward.

Significance of the Research on China's S&T Roadmap to 2050

We are aware that the National Mid- and Long-term S&T Plan to 2020 has already been formed after two years' hard work by a panel of over 2000 experts and scholars brought together from all over China, chaired by Premier Wen Jiabao. This clearly shows that China has already had its S&T blueprint to 2020. Then, why did CAS conduct this research on China's S&T roadmap to 2050?

In the summer of 2007 when CAS was working out its future strategic priorities for S&T development, it realized that some issues, such as energy, must be addressed with a long-term view. As a matter of fact, some strategic researches have been conducted, over the last 15 years, on energy, but mainly on how to best use of coal, how to best exploit both domestic and international oil and gas resources, and how to develop nuclear energy in a discreet way. Renewable energy was, of course, included but only as a supplementary energy. It was not yet thought as a supporting leg for future energy development. However, greenhouse gas emissions are becoming a major world concern over

* It is adapted from a speech by President Yongxiang Lu at the first High-level Workshop on China's S&T Roadmap for Priority Areas to 2050, organized by the Chinese Academy of Sciences, in October, 2007.

the years, and how to address the global climate change has been on the agenda. In fact, what is really behind is the concern for energy structure, which makes us realize that fossil energy must be used cleanly and efficiently in order to reduce its impact on the environment. However, fossil energy is, pessimistically speaking, expected to be used up within about 100 years, or optimistically speaking, within about 200 years. Oil and gas resources may be among the first to be exhausted, and then coal resources follow. When this happens, human beings will have to refer to renewable energy as its major energy, while nuclear energy as a supplementary one. Under this situation, governments of the world are taking preparatory efforts in this regard, with Europe taking the lead and the USA shifting to take a more positive attitude, as evidenced in that: while fossil energy has been taken the best use of, renewable energy has been greatly developed, and the R&D of advanced nuclear energy has been reinforced with the objective of being eventually transformed into renewable energy. The process may last 50 to 100 years or so. Hence, many S&T problems may come around. In the field of basic research, for example, research will be conducted by physicists, chemists and biologists on the new generation of photovoltaic cell, dye-sensitized solar cells (DSC), high-efficient photochemical catalysis and storage, and efficient photosynthetic species, or high-efficient photosynthetic species produced by gene engineering which are free from land and water demands compared with food and oil crops, and can be grown on hillside, saline lands and semi-arid places, producing the energy that fits humanity. In the meantime, although the existing energy system is comparatively stable, future energy structure is likely to change into an unstable system. Presumably, dispersive energy system as well as higher-efficient direct current transmission and storage technology will be developed, so will be the safe and reliable control of network, and the capture, storage, transfer and use of CO_2, all of which involve S&T problems in almost all scientific disciplines. Therefore, it is natural that energy problems may bring out both basic and applied research, and may eventually lead to comprehensive structural changes. And this may last for 50 to 100 years or so. Taking the nuclear energy as an example, it usually takes about 20 years or more from its initial plan to key technology breakthroughs, so does the subsequent massive application and commercialization. If we lose the opportunity to make foresighted arrangements, we will be lagging far behind in the future. France has already worked out the roadmap to 2040 and 2050 respectively for the development of the 3rd and 4th generation of nuclear fission reactors, while China has not yet taken any serious actions. Under this circumstance, it is now time for CAS to take the issue seriously, for the sake of national interests, and to start conducting a foresighted research in this regard.

This strategic research covers over some dozens of areas with a long-term view. Taking agriculture as an example, our concern used to be limited only to the increased production of high-quality food grains and agricultural by-products. However, in the future, the main concern will definitely be given to the water-saving and ecological agriculture. As China is vast in territory, diversified technologies in this regard are the appropriate solutions. Animal husbandry has been used by developed countries, such as Japan and Denmark, to make bioreactor and pesticide as well. Plants have been used by Japan to make bioreactors which are safer and cost-effective than that made from animals. Potato, strawberry, tomato and the like have been bred in germ-free greenhouses, and value-added products have been made through gene transplantation technology. Agriculture in China must not only address the food demands from its one billions-plus population, but also take into consideration of the value-added agriculture by-products and the high-tech development of agriculture as well. Agriculture in the future is expected to bring out some energies and fuels needed by both industry and man's livelihood as well. Some developed countries have taken an earlier start to conduct foresighted research in this regard, while we have not yet taken sufficient consideration.

Population is another problem. It will be most likely that China's population will not drop to about 1 billion until the end of this century, given that the past mistakes of China's population policy be rectified. But the subsequent problem of ageing could only be sorted out until the next century. The current population and health policies face many challenges, such as, how to ensure that the 1.3 to 1.5 billion people enjoy fair and basic public healthcare; the necessity to develop advanced and public healthcare and treatment technologies; and the change of research priority to chronic diseases from infectious diseases, as developed countries have already started research in this regard under the increasing social and environmental change. There are many such research problems yet to be sorted out by starting from the basic research, and subsequent policies within the next 50 years are in need to be worked out.

Space and oceans provide humanity with important resources for future development. In terms of space research, the well-known Manned Spacecraft Program and China's Lunar Exploration Program will last for 20 or 25 years. But what will be the whole plan for China's space technology? What is the objective? Will it just follow the suit of developed countries? It is worth doing serious study in this regard. The present spacecraft is mainly sent into space with chemical fuel propellant rocket. Will this traditional propellant still be used in future deep space exploration? Or other new technologies such as electrical propellant, nuclear energy propellant, and solar sail technologies be developed? We haven't yet done any strategic research over these issues, not even worked out any plans. The ocean is abundant in mineral resources, oil and gas, natural gas hydrate, biological resources, energy and photo-free biological evolution, which may arise our scientific interests. At present, many countries have worked out new strategic marine plans. Russia, Canada, the USA, Sweden and Norway have centered their contention upon the North Pole, an area of strategic significance. For this, however, we have only limited plans.

The national and public security develops with time, and covers both conventional and non-conventional security. Conventional security threats only refer to foreign invasion and warfare, while, the present security threat may come out from any of the natural, man-made, external, interior, ecological, environmental, and the emerging networking (including both real and virtual) factors. The conflicts out of these must be analyzed from the perspective of human civilization, and be sorted out in a scientific manner. Efforts must be made to root out the cause of the threats, while human life must be treasured at any time.

In general, it is necessary to conduct this strategic research in view of the future development of China and mankind as well. The past 250 years' industrialization has resulted in the modernization and better-off life of less than 1 billion people, predominantly in Europe, North America, Japan and Singapore. The next 50 years' modernization drive will definitely lead to a better-off life for 2–3 billion people, including over 1 billion Chinese, doubling or tripling the economic increase over that of the past 250 years, which will, on the one hand, bring vigor and vitality to the world, and, on the other hand, inevitably challenge the limited resources and eco-environment on the earth. New development mode must be shaped so that everyone on the earth will be able to enjoy fairly the achievements of modern civilization. Achieving this requires us, in the process of China's modernization, to have a foresighted overview on the future development of world science and human civilization, and on how science and technology could serve the modernization drive. S&T roadmap for priority areas to 2050 must be worked out, and solutions to core science problems and key technology problems must be straightened out, which will eventually provide consultations for the nation's S&T decision-making.

Possibility of Working out China's S&T Roadmap to 2050

Some people held the view that science is hard to be predicted as it happens unexpectedly and mainly comes out of scientists' innovative thinking, while, technology might be predicted but at the maximum of 15 years. In my view, however, S&T foresight in some areas seems feasible. For instance, with the exhaustion of fossil energy, some smart people may think of transforming solar energy into energy-intensive biomass through improved high-efficient solar thin-film materials and devices, or even developing new substitute. As is driven by huge demands, many investments will go to this emerging area. It is, therefore, able to predict that, in the next 50 years, some breakthroughs will undoubtedly be made in the areas of renewable energy and nuclear energy as well. In terms of solar energy, for example, the improvement of photoelectric conversion efficiency and photothermal conversion efficiency will be the focus. Of course, the concrete technological solutions may be varied, for example, by changing the morphology of the surface of solar cells and through the reflection, the entire spectrum can be absorbed more efficiently; by developing multi-layer functional thin-films for transmission and absorption; or by introducing of nanotechnology and quantum control technology, *etc.* Quantum control research used to limit mainly to the solution to information functional materials. This is surely too narrow. In the future, this research is expected to be extended to the energy issue or energy-based basic research in cutting-edge areas.

In terms of computing science, we must be confident to forecast its future development instead of simply following suit as we used to. This is a possibility rather than wild fancies. Information scientists, physicists and biologists could be engaged in the forward-looking research. In 2007, the Nobel Physics Prize was awarded to the discovery of colossal magneto-resistance, which was, however, made some 20 years ago. Today, this technology has already been applied to hard disk store. Our conclusion made, at this stage, is that: it is possible to make long-term and unconventional S&T predictions, and so is it to work out China's S&T roadmap in view of long-term strategies, for example, by 2020 as the first step, by 2030 or 2035 as the second step, and by 2050 as the maximum.

This possibility may also apply to other areas of research. The point is to emancipate the mind and respect objective laws rather than indulging in wild fancies. We attribute our success today to the guidelines of emancipating the mind and seeking the truth from the facts set by the Third Plenary Session of the 11th Central Committee of the Communist Party of China in 1979. We must break the conventional barriers and find a way of development fitting into China's reality. The history of science tells us that discoveries and breakthroughs could only be made when you open up your mind, break the conventional barriers, and make foresighted plans. Top-down guidance on research with increased financial support and involvement of a wider range of talented scientists is not in conflict with demand-driven research and free discovery of science as well.

Necessity of CAS Research on China's S&T Roadmap to 2050

Why does CAS launch this research? As is known, CAS is the nation's highest academic institution in natural sciences. It targets at making basic, forward-looking and strategic research and playing a leading role in China's science. As such, how can it achieve this if without a foresighted view on science and technology? From the perspective of CAS, it is obligatory to think, with a global view, about what to do after the 3rd Phase of the Knowledge Innovation Program (KIP). Shall we follow the way as it used to? Or shall we, with a view of national interests, present our in-depth insights into different research disciplines, and make efforts to reform the organizational structure and system, so that the innovation capability of CAS and the nation's science and technology mission will be raised to a new height? Clearly, the latter is more positive. World science and technology develops at a lightening speed. As global economy grows, we are aware that we will be lagging far behind if without making progress, and will lose the opportunity if without making foresighted plans. S&T innovation requires us to make joint efforts, break the conventional barriers and emancipate the mind. This is also what we need for further development.

The roadmap must be targeted at the national level so that the strategic research reports will form an important part of the national long-term program. CAS may not be able to fulfill all the objectives in the reports. However, it can select what is able to do and make foresighted plans, which will eventually help shape the post-2010 research priorities of CAS and the guidelines for its future reform.

Once the long-term roadmap and its objectives are identified, system mechanism, human resources, funding and allocation should be ensured for full implementation. We will make further studies to figure out: What will happen to world innovation system within the next 30 to 50 years? Will universities, research institutions and enterprises still be included in the system? Will research institutes become grid structure? When the cutting-edge research combines basic science and high-tech and the transformative research integrates the cutting-edge research with industrialization, will that be the research trend in some disciplines? What will be the changes for personnel structure, motivation mechanism and upgrading mechanism within the innovation system? Will there be any changes for the input and structure of innovation resources? If we could have a clear mind of all the questions, make foresighted plans and then dare to try out in relevant CAS institutes, we will be able to pave a way for a more competitive and smooth development.

Social changes are without limit, so are the development of science and technology, and innovation system and management as well. CAS must keep moving ahead to make foresighted plans not only for science and technology, but also for its organizational structure, human resources, management modes, and resource structures. By doing so, CAS will keep standing at the forefront of science and playing a leading role in the national innovation system, and even, frankly speaking, taking the lead in some research disciplines in the world. This is, in fact, our purpose of conducting the strategic research on China's S&T roadmap.

Prof. Dr.-Ing. Yongxiang Lu

President of the Chinese Academy of Sciences

Preface to the Roadmaps 2050

CAS is the nation's think tank for science. Its major responsibility is to provide S&T consultations for the nation's decision-makings and to take the lead in the nation's S&T development.

In July, 2007, President Yongxiang Lu made the following remarks: "In order to carry out the Scientific Outlook of Development through innovation, further strategic research should be done to lay out a S&T roadmap for the next 20–30 years and key S&T innovation disciplines. And relevant workshops should be organized with the participation of scientists both within CAS and outside to further discuss the research priorities and objectives. We should no longer confine ourselves to the free discovery of science, the quantity and quality of scientific papers, nor should we satisfy ourselves simply with the Principal Investigators system of research. Research should be conducted to address the needs of both the nation and society, in particular, the continued growth of economy and national competitiveness, the development of social harmony, and the sustainability between man and nature. "

According to the Executive Management Committee of CAS in July, 2007, CAS strategic research on S&T roadmap for future development should be conducted to orchestrate the needs of both the nation and society, and target at the three objectives: the growth of economy and national competitiveness, the development of social harmony, and the sustainability between man and nature.

In August, 2007, President Yongxiang Lu further put it: "Strategic research requires a forward-looking view over the world, China, and science & technology in 2050. Firstly, in terms of the world in 2050, we should be able to study the perspectives of economy, society, national security, eco-environment, and science & technology, specifically in such scientific disciplines as energy, resources, population, health, information, security, eco-environment, space and oceans. And we should be aware of where the opportunities and challenges lie. Secondly, in terms of China's economy and society in 2050, we should take into consideration of factors like: objectives, methods, and scientific supports needed for economic structure, social development, energy structure, population and health, eco-environment, national security and innovation capability. Thirdly, in terms of the guidance of Scientific Outlook of Development on science and technology, it emphasizes the people's interests and development, science and technology, science and economy, science and society, science and eco-environment, science and culture, innovation and collaborative development. Fourthly, in terms of the supporting role of research in scientific development, this includes how to optimize the economic structure and boost economy, agricultural development, energy structure, resource conservation, recycling economy, knowledge-based society, harmonious coexistence between man and nature, balance of regional development, social harmony, national security, and international cooperation. Based on these, the role of CAS will be further identified."

Subsequently, CAS launched its strategic research on the roadmap for priority areas to 2050, which comes into eighteen categories including: energy, water resources, mineral resources, marine resources, oil and gas, population and health, agriculture, eco-environment, biomass resources, regional development, space, information, advanced manufacturing, advanced materials, nano-science, big science facilities, cross-disciplinary and frontier research, and national and public security. Over 300 CAS experts in science, technology, management and documentation & information, including about 60 CAS members, from over 80 CAS institutes joined this research.

Over one year's hard work, substantial progress has been made in each research group of the scientific disciplines. The strategic demands on priority areas in China's modernization drive to 2050 have been strengthened out; some core science problems and key technology problems been set forth; a relevant S&T roadmap been worked out based on China's reality; and eventually the strategic reports on China's S&T roadmap for eighteen priority areas to 2050 been formed. Under the circumstance, both the Editorial Committee and Writing Group, chaired by President Yongxiang Lu, have finalized the general report. The research reports are to be published in the form of CAS strategic research serial reports, entitled *Science and Technology Roadmap to China 2050: Strategic Reports of the Chinese Academy of Sciences.*

The unique feature of this strategic research is its use of S&T roadmap approach. S&T roadmap differs from the commonly used planning and technology foresight in that it includes science and technology needed for the future, the roadmap to reach the objectives, description of environmental changes, research needs, technology trends, and innovation and technology development. Scientific planning in the form of roadmap will have a clearer scientific objective, form closer links with the market, projects selected be more interactive and systematic, the solutions to the objective be defined, and the plan be more feasible. In addition, by drawing from both the foreign experience on roadmap research and domestic experience on strategic planning, we have formed our own ways of making S&T roadmap in priority areas as follows:

(1) Establishment of organization mechanism for strategic research on S&T roadmap for priority areas

The Editorial Committee is set up with the head of President Yongxiang Lu and the involvement of Chunli Bai, Erwei Shi, Xin Fang, Zhigang Li, Xiaoye Cao and Jiaofeng Pan. And the Writing Group was organized to take responsibility of the research and writing of the general report. CAS Bureau of Planning and Strategy, as the executive unit, coordinates the research, selects the scholars, identifies concrete steps and task requirements, sets forth research approaches, and organizes workshops and independent peer reviews of the research, in order to ensure the smooth progress of the strategic research on the S&T roadmap for priority areas.

(2) Setting up principles for the S&T roadmap for priority areas

The framework of roadmap research should be targeted at the national level, and divided into three steps as immediate-term (by 2020), mid-term (by 2030) and long-term (by 2050). It should cover the description of job requirements, objectives, specific tasks, research approaches, and highlight core science problems and key technology problems, which must be, in general, directional, strategic and feasible.

(3) Selection of expertise for strategic research on the S&T roadmap

Scholars in science policy, management, information and documentation, and chief scientists of the middle-aged and the young should be selected to form a special research group. The head of the group should be an outstanding scientist with a strategic vision, strong sense of responsibility and coordinative capability. In order to steer the research direction, chief scientists should be selected as the core members of the group to ensure that the strategic research in priority areas be based on the cutting-edge and frontier research. Information and documentation scholars should be engaged in each research group to guarantee the efficiency and systematization of the research through data collection and analysis. Science policy scholars should focus on the strategic demands and their feasibility.

(4) Organization of regular workshops at different levels

Workshops should be held as a leverage to identify concrete research steps and ensure its smooth progress. Five workshops have been organized consecutively in the following forms:

High-level workshop on S&T strategies. Three workshops on S&T strategies have been organized in October, 2007, December, 2007, and June, 2008, respectively, with the participation of research group heads in eighteen priority areas, chief scholars, and relevant top CAS management members. Information has been exchanged, and consensus been reached to ensure research directions. During the workshops, President Yongxiang Lu pinpointed the significance, necessity and possibility of the roadmap research, and commented on the work of each research groups, thus pushing the research forward.

Special workshops. The Editorial Committee invited science policy scholars to the special workshops to discuss the eight basic and strategic systems for China's socio-economic development. Perspectives on China's science-driven modernization to 2050 and characteristics and objectives of the eight systems have been outlined, and twenty-two strategic S&T problems affecting the modernization have been figured out.

Research group workshops. Each research group was further divided into different research teams based on different disciplines. Group discussions, team discussions and cross-team discussions were organized for further research, occasionally with the involvement of related scholars in special topic discussions. Research group workshops have been held some 70 times.

Cross-group workshops. Cross-group and cross-disciplinary workshops were organized, with the initiation by relative research groups and coordination by Bureau of Planning and Strategies, to coordinate the research in relative disciplines.

Professional workshops. These workshops were held to have the suggestions and advices of both domestic and international professionals over the development and strategies in related disciplines.

(5) Establishment of a peer review mechanism for the roadmap research

To ensure the quality of research reports and enhance coordination among different disciplines, a workshop on the peer review of strategic research on the S&T roadmap was organized by CAS Bureau of Planning and Strategy, in November, 2008, bringing together of about 30 peer review experts and 50 research group scholars. The review was made in four different categories, namely, resources and environment, strategic high-technology, bio-science & technology, and basic research. Experts listened to the reports of different research groups, commented on the general structure, what's new and existing problems, and presented their suggestions and advices. The outcomes were put in the written forms and returned to the research groups for further revisions.

(6) Establishment of a sustained mechanism for the roadmap research

To cope with the rapid change of world science and technology and national demands, a roadmap is, by nature, in need of sustained study, and should be revised once in every 3–5 years. Therefore, a panel of science policy scholars should be formed to keep a constant watch on the priority areas and key S&T problems for the nation's long-term benefits and make further study in this regard. And hopefully, more science policy scholars will be trained out of the research process.

The serial reports by CAS have their contents firmly based on China's reality while keeping the future in view. The work is a crystallization of the scholars' wisdom, written in a careful and scrupulous manner. Herewith, our sincere gratitude goes to all the scholars engaged in the research, consultation and review. It is their joint efforts and hard work that help to enable the serial reports to be published for the public within only one year.

To precisely predict the future is extremely challenging. This strategic research covered a wide range of areas and time, and adopted new research approaches. As such, the serial reports may have its deficiency due to the limit in knowledge and assessment. We, therefore, welcome timely advice and enlightening remarks from a much wider circle of scholars around the world.

The publication of the serial reports is a new start instead of the end of the strategic research. With this, we will further our research in this regard, duly release the research results, and have the roadmap revised every five years, in an effort to provide consultations to the state decision-makers in science, and give suggestions to science policy departments, research institutions, enterprises, and universities for their S&T policy-making. Raising the public awareness of science and technology is of great significance for China's modernization.

Writing Group of the General Report

February, 2009

Preface

Study of regional development is a cross-disciplinary field centered on economic geography, while also involving adjacent sub-disciplines such as physical geography, natural resources, environment, ecology, development economics, *etc*. It focuses on the analysis of both economic and social factors underlying the formation of territorial spatial structure in various forms.

Regional development of China is characterized by complex conditions and diversified development types. Such complexity is especially reflected in the emerging problems of disorderly territorial spatial structure and serious challenges confronting regional sustainable development in current era of rapid industrialization and urbanization. Regional development problem has become one of the core issues for policy-makers at all levels as well as a major practical and theoretical question that draws the attention of both academic scholars and the public. In the next three to five decades, China's regional development will be confronted with severe challenges brought about by global climate change and economic globalization. The increase of population and urbanization level will also put regional development study under severe pressure to meet imminent demand of reconstructing man-earth areal system. Research on the scientific and technological development roadmap in China's regional development studies during the additional years to 2050 is a strategic and far-insighted deployment. The research findings of such roadmap are expected to play a vital role in the improvement of regional competitiveness and sustainable development capacity and the promotion of coordinated regional development.

Research on the roadmap of regional development is mainly carried out by the Institute of Geographic Sciences and Natural Resources Research, CAS (Chinese Academy of Sciences) with the joint efforts of the Nanjing Institute of Geography and Limnology, the Northeast Institute of Geography and Agroecology, the Chengdu Institute of Mountain Hazards and Environment and the Xinjiang Institute of Ecology and Geography, *etc*. Since the 1950s when CAS was firstly founded, the research team of regional development had been oriented toward meeting national strategic demand and resolving key issues of regional development with a core focus on the coordinated development of resources, environment and social economy, leading to the formation of a distinct research field in CAS. Especially since the implementation of the Knowledge Innovation Program (KIP) in CAS, the team has continuously published *China's Regional Development Reports*, successfully completed many important regional planning projects such as "Major Function Oriented

Zoning", "planning for the old industrial bases in Northeast China", and "the metropolitan area of Beijing-Tianjin-Hebei", *etc*. Furthermore, the team has also submitted to the central government a number of important consultation reports on the adjustment of urbanization development and new rural construction strategies, and has produced a system of theories concerning man-earth areal system and territorial development. Regional development research team of CAS has increasingly become a leading academic center in China studies and a "think tank" supporting central policy-making and planning in regional development.

Based on the features of regional development research and the analysis of the scientific and technological demand of China's future regional development, the research on the roadmap of regional development has made some innovations in its research framework and organizational structure in addition to following the general arrangement by CAS. Firstly, we adopt an integrated structure with a mixture of comprehensive report with general summary, thematic report focused on key industries and areas of regional development, and regional report focused on development problems in representative regions. Secondly, the research team is diversified in age structure as it includes both experienced senior experts and young scholars. Thirdly, scholars both inside and outside the CAS, as well as policy-makers are all involved in the research. Therefore, we benefit a lot from the suggestions and comments made by scholars in various academic institutions including Development Research Center of the State Council, Chinese Academy of Social Sciences as well as those by officials in central and local governments. In this sense, the publication of this book is the joint efforts of numerous institutions and people, to which we would like to extend our sincerest thanks.

The roadmap of regional development research systematically discusses the role of regional development study in contemporary science system, its features, scientific demand out of future national and societal development, major research questions, research orientation, methodology, supporting platform *etc*. Specifically, it takes the strengthening of research on "man-earth areal system" dynamics or environment-society dynamics as the direction of theoretical development in this field, and aims to build "Simulation and Decision Support System for Regional Sustainable Development in China" in practical work for improving the research capabilities in regional development. The roadmap is not only a research report, but also a long-term planning in the scientific field. We sincerely hope that our work may draw the attention of scholars in relevant fields both at home and abroad.

Research Group on Regional Development
Research of the Chinese Academy of Sciences
March, 2010

Contents

Abstract

Regional development plays a vital role in the modernization of all countries, and is of great significance in the implementation of the Scientific Outlook on Development and the construction of harmonious society in China. In the next 30 to 50 years, global climate change and economic globalization will have increasingly substantial influence on regional development in China and regional response and reaction will be an integrated choice to adapt to the restriction of carbon emission reduction and the supporting conditions of "two types of resources and two markets". At that time, the population of China will be over 1.5 billion, the 70% of which will live in urban areas. As a result, the land demand of food security, ecological security and urbanization will continue to increase, and urbanization and regional development will be confronted with a core task of continuously adjusting and restructuring the interactive relationship between human and the natural system. Under the combined influence of state power and market force, the spatial agglomeration and dispersion processes of population and industries will be intensified significantly, and the spatial pattern of regional development based on the division of territorial function will be formed. Regional development will have increasingly enriched contents and multiple objectives. The interaction between regions at different spatial scales will tend to be diversified.

Regional development research is a burgeoning interdisciplinary field in China with distinctive features of comprehensiveness, spatial diversity, time variability and uncertainty. It contains 6 major modules: modern methods and techniques in regional analysis, which is the methodological basis of the whole research; the affecting factors of regional development process and their driving mechanisms; theories on the distribution of different social and economic sectors;the spatial organization principles of different types of regions; theories on the formation and spatial-temporal evolution of regional development pattern; the policy system for regulating human spatial behavior and territorial function.

Based on the analysis of regional development tendency and future scientific demand, the major tasks in the short run (up to 2020) are to reveal regional response and spatial process under economic globalization and global climate change, to improve the spatial recognition system of territorial function, to develop theories and methods on the processes of inter-regional material and energy flow, the agglomeration and dispersion of population and industries as well as ecological compensation and financial transfer payment, in order to build a data collection network and an information platform for the support of regional development simulation. In the mid-term period (up to 2030), the major tasks are to understand the physical basis, material ensurance, economic and social process and environmental effects of regional sustainable development, to discuss the spatial pattern of economic growth and social development in different development phases in the future, to investigate the regional development mechanism and tendency in the "three-dimensional objective space of economy-society-ecosystem" and the scientific process and model of urbanization in China, to construct function zones in conformity with the principle of coordinated regional development and the underlying regional planning and policy support system. The main objective is to develop different types of regional sustainable development models under the support of spatial analysis and simulation technique and decision-making support system for the guidance of practical work. In the long run (up to 2050), the major tasks are to explore the interactive mechanism between environment and development and between their various elements, to reveal the structural features and evolutionary principles of man-earth areal system, to develop the theoretical system of man-earth and environment-society dynamics for wide application in the simulation and decision-making support system of urbanization and regional development, to establish optimal land security pattern on the basis of ecological security pattern at different spatial scales, and to carry out large-scale empirical researches on cross-border regional cooperation and development issues.

The Chinese Academy of Sciences will strengthen its function as "think tank" in central formulation of regional development strategy through the development of regional development theories, methods, data and technical platform. In addition, it will build the "Simulation and Decision Support System for Regional Sustainable Development in China" and a world-class research base of regional sustainable development based on the combination of long-term research plan and short-term research project.

1 Regional Development and Regional Development Research

Regional development is an important interdisciplinary research field among geography, resource and environmental sciences, economics and management sciences that is characterized by three distinctive features of comprehensiveness, spatial diversity and time variability[1]. Due to the extremely complex role of "human" in its affecting factors, driving mechanism and development status, regional development displays an eminent feature of uncertainty. It is not only the important foundation of modernization process in China, but also a key dimension of the implementation of the Scientific Outlook on Development and the construction of harmonious society. The carrying capacity of natural resources and eco-environmental foundation is weak with great disparities in the natural background and physical structure of different regions in China. Meanwhile, the pressure exerted by the increasing amount of economic aggregate and the expanding process of urbanization is intensified. As a result, a large number of scientific and technologic issues in regional development are of great significance to the development of our country which is characterized by such distinctive features and is currently at the stage of industrialization[2,3]. Internationally, there have been some scholars in developed countries conducting research on land and regional development and getting involved in the making of territorial governance policies by various governments[4,5]. In fact, some key international programs, such as the "International Human Dimensions Program" (IHDP) has already revealed the interactive relationship between human socio-economic activities and physical foundation at different spatial scales through a combination of socio-economic development with global changes in natural sciences [6].

The Epochal Background of China's Regional Development in the Next 30 Years

(1) Global climate change and economic globalization will have more substantial influence on regional development in China. Regional response and reaction will be an integrated choice to adapt to the restriction of carbon emission reduction and the supporting conditions of "two types of resources and two markets".

(2) The population of China will be over 1.5 billion(Hong Kong,Macao and Taiwan data not available,sic passim), the 70% of which will live in urban areas. As a result, the land demand of food security, ecological ensurance and urbanization will continue to increase, and, urbanization and regional development will be confronted with a core task of continuously adjusting, optimizing and restructuring the interactive relationship between human and the natural system.

(3) Under the combined influence of state power and market force, the spatial agglomeration and dispersion process of population and industries will be intensified significantly, and the spatial pattern of regional development based on the division of territorial function will be formed.

(4) Regional development will have increasingly enriched contents and multiple objectives. The interaction between regions at different spatial scales will tend to be diversified.

1.1 The Attribute of Regional Development Research

Regional economy in the past 50 years has developed rapidly all over the world. While the application of strong social force and productive technologies has produced material and spiritual civilization at significant scale, it has simultaneously reshaped the socio-economic and natural structures at global, national and regional scales. Since the reform and opening up of China, regional development has been one of the core issues of central policy-making. The scientific and technological innovation in regional development has not only played an important role in coordinating socio-economic development and resources and environment, but also effectively promoted the disciplinary construction of regional research[7].

1.1.1 The Comprehensiveness of Regional Development and Integrated Research Methods

Regional development issue is a complex and giant system composed of both human and natural phenomena on the earth surface. It thus displays remarkable feature of comprehensiveness. The affecting factors of regional development are comprehensive, including human factors such as economy and society, natural factors such as environment and natural resources, and many other complex factors with undefinable nature. The driving mechanisms of regional development include not only basic theories in such humanities as economics, social science and cultural study, but also natural processes in such fields as physics, chemistry and biology. In certain cases, some driving forces exhibit complex mechanisms under the combined influence of all above processes. Moreover, not only different factors have influence on the development of different regions in different ways and degrees, but also the same factor may have different influences on the development of the same region at different phases[8].

Regional development research is a comprehensive and interdisciplinary field with the attributes of both natural sciences and social sciences. As the carrier of various human and natural phenomena on the earth surface, region is characterized by the interweaving of a variety of problems which need to be solved through comprehensive perspective. Therefore, comprehensiveness, which refers to the method of analyzing regional development with a systematic and holistic view, is always the top-level research question in regional development. Such comprehensiveness is reflected not only in research on driving mechanisms of key physical-geographical processes that pay increasing attention to the effect of human activity such as global climate change, but also in research on human and economic geographical activities that increasingly take resource and environment into account. Both factor and regional analyses are major methods of integrated research on regional development[9,10].

Factor Method in the Comprehensive Research of Regional Development

The factor method focuses on various factors in the earth surface system and aims to reveal the structures, functions and spatial-temporal evolution theories of the earth surface system and its subsystems by probing into the factors themselves and the interactive relationship among them. More often than not, the factor method fails to reach its policy objectives as a result of its partial understanding of the affecting factors and their mutual interactions. At present, the factor method is mainly employed in two types of research in China. One is to emphasize the effect of natural environment on social and economic development, including the evaluation of development status and the selection of development objective. For example, increasingly attention has been paid to reasonable use of natural resources, disaster prevention and reduction and environment conservation, natural environment constraints on economic growth, as well as the minimization of damage to ecological environment brought about by economic development process in land development and regional planning. Such change is due to the recognition of resource, environment and disaster attributes of natural environment [6]. Therefore, the effects of natural environment on social and economic development should be comprehensively taken into consideration. The other type of research is to identify the effects of human activities in key physical-geographical process. For example, the impact of human activities is highly valued in carbon cycle study through detailed analysis of the mutual interaction among land use structure, energy use structure, industrial structure and carbon cycle, and tentative investigation of the possible effect of international trade and industrial transfer on carbon cycle policy, from the perspective of greenhouse gases emission (mainly methane and carbon dioxide). However, there is a major flaw in such exploratory research to treat human activity factors or environmental factors independently without probing into the intrinsic mechanism of the system as an organic whole.

Regional Method in the Comprehensive Research of Regional Development

As both the starting and ending point of regional development research, region is characterized by the interweaving of a variety of problems in need of comprehensive solution. Therefore, the integrated researches on regional development are mostly reflected in regional research and most of the key regional research programs at different levels tend to be comprehensive. For example, the research focus of Tibetan Plateau has been extended from its uplift mechanism and environmental effects to the integrated study of both human and natural factors in regional sustainable development. Similarly, the research on the Loess Plateau and the karst region in southwest China is also a case in point in which research focus has been shifted from physical-geographical perspective of water and soil erosion prevention to more comprehensive treatment of both prevention technique and regional development policy. Moreover, the research on the arid region in northwest China has always been emphasizing regional sustainable development with equal importance attached to natural and human factors. In addition to these macro-scale comprehensive programs, some meso-and micro-scale researches of regions with intense interaction between human and nature elements also tend to be comprehensive, such as research on mining areas which focuses on the coupling relationship between the velocity and intensity of the transformation from natural landscape to human landscape and the shift of natural-environmental system, and research on the theory of landscape recovery. For urban-rural transient zone, the core problem of research is the ecological and environmental effects of internal transformation of human landscape, while the core problem of research on ecologically fragile zone in agricultural and pastoral areas concerns the establishment of evaluation measurement system of the comprehensive benefits of landscape counter-transformation. As for industrialized region in rural areas, focus should be placed on the investigation of the ecological and environmental effects amidst the disorderly and complex process of human landscape evolution.

1.1.2 General Theories of Spatial Differentiation and Regional Development

Spatial differentiation is an inevitable phenomenon emanating from regional development process because of the spatially imbalanced distribution of both natural and human factors affecting the patterns, processes, structures and functions of man-earth areal system. The same affecting factor may display different effects on regional development in different regions or at different development stages. Even the same driving force may have varied effects at different spatial scales. Therefore, the imbalanced distribution of affecting factors and the spatial differentiation of their driving mechanisms are the fundamental causes of regional disparities[11].

Regional differentiation is not only a distinct feature but also a serious problem of regional development in our country. China has a vast territory with significant areal differentiations in natural conditions. For a long time, spatial differentiations in natural conditions characterized by the division of three natural areas based on climate and elevation has defined the physical-geographical environment underlying economic development in different regions. Such spatial differentiation, together with spatial disparities in location, internal and external linkages, and technological base, contributes to the formation of significant and complex differentiation in regional development. The solving of regional development problems entails the exploration of the formation and evolutionary mechanism of regional differentiation on the basis of a full understanding of the differentiation of man-earth areal system and consequently optimal organization of spatial structure. Regional development research is exactly the discipline with significant feature of territoriality which aims to reveal regional differentiation in terms of internal structure, inter-regional relationship, and the role of region in the whole system. Therefore, regional development research is conducive to the grasp of regional differentiation theory and the exploration of rational path toward coordinated regional development.

Spatial differentiation is the basic theory of regional development, and should be taken as the basis of regional governance. However, many of our regional development policies in the past were made without a full understanding of the basic theory of regional differentiation. In recent years, China has put forward many important regional development strategies such as the Development of Western China, Revitalization of Traditional Industrial Bases in Northeast China, the Rise of Central China and the Prioritized Development of East China, *etc*. While All these strategies formulated on the basis of macro region have made certain contribution to the modernization process of our country, they have failed to solve the problems of disorderly structure of regional development, deterioration of ecological environment, increase of resource pressure and expanding regional disparities that are brought about by high-speed industrialization and urbanization due to inadequate knowledge of regional differentiation. In the long run, such ignorance may hinder the construction of a well-off society in all round ways, the realization of modernization objectives, and the improvement of international competitiveness of our country. Therefore, in the new period of development, it is necessary for China to have a correct and comprehensive understanding of regional differentiation in order to make rational decisions over regional development.

Existing literature on the general theories of regional development mainly focus on three aspects, namely (Fig.1.1): internal development problems of regions at different spatial scales including the coupling process between social-economic development and natural resource and environment (A or B); spatial differentiation theories and interaction among regions on the same spatial scale (A and B); the problems occurring in the transformation process of different spatial patterns and scales (C).

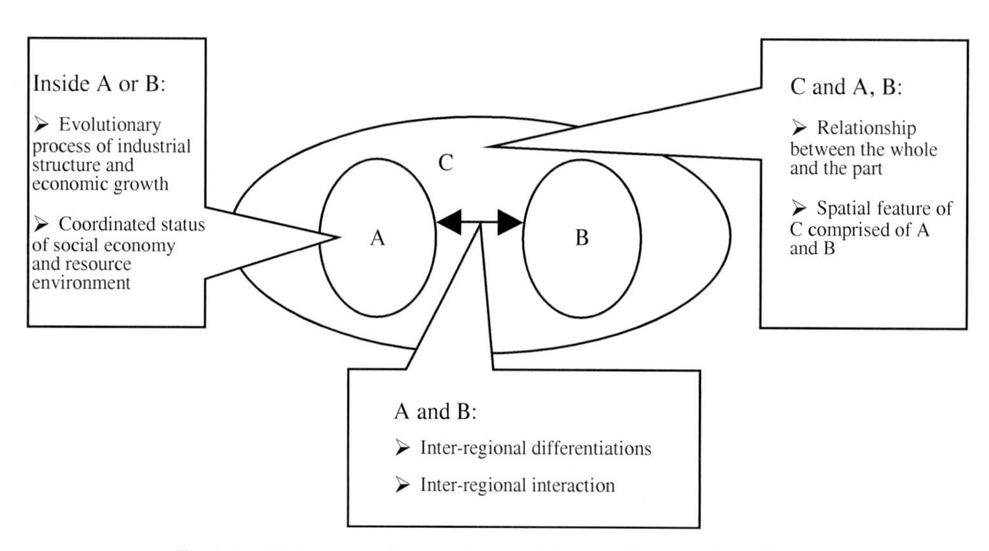

Fig.1.1 Major research area of general theory of regional development

1.1.3 New Factors, New Mechanisms and Time Differentiation in Regional Development

Regional development is a progress of earth surface change in a certain region under the combined influence of various factors. The direction and extent of change of these affecting factors may vary at different development stages. Meanwhile, the mechanisms of regional development, *i.e.* the mutual interaction among various elements in man-earth areal system, may also be different at different development phases[12].

The continuous emergence of new affecting factors and new mechanisms is one of the important causes of time differentiation in regional development. On the one hand, the appearance of new affecting factors and mechanisms of regional development is the result of social evolution. There are different affecting factors at different development stages during the development history of humankind. At the stage of agricultural society, it is natural resources, such as climate, water and land resources, that have fundamental impact on regional development, while industrial raw materials, such as coal, iron, oil, natural gas, non-ferrous metals and non-metallic mineral resources, became the dominant factors of regional development at the stage of industrial society. All of them are traditional affecting factors of regional development. Recently, some factors such as informatization, globalization, human capital, and institutions began to emerge as new leading factors of regional development with the advent of the post-industrial society.

On the other hand, new factors and mechanisms are also the outcome of societal transformation. Regional development was once considered to be a pure economic behavior, and the process of regional development was understood simply as industrialization process. As a result, many early theories of regional development were based on economic growth and profit maximization. However, with the ever accumulation of both theoretical and practical knowledge, more and more people began to realize that regional development process is not only a simple economic behavior, but also a social behavior in which any impact on natural and ecological system will be fed back to the long-term development of economic system itself. Therefore, contemporary regional development should pay attention to its integrated process and overall interests. It is obvious that regional development process is not necessarily affected by economic ones only but non-economic factors such as social values, environmental ethics, ecological civilization, *etc.* as well. While these factors were not valued in traditional theories of regional development, they may become new affecting factors of regional development.

Since the 1970s, the nature of regional development has been extended to a series of economic and social processes, involving continuously income increase, technological improvement, industrial upgrading, expansion of external linkage, the demand structure change, and institutional transformation, *etc.*, owing much to stagnant economic growth, prominent structural contradictions, serious social problems and deteriorating natural environment.

As a consequence, some new factors, particularly human factors related to human activities, such as informatization, internationalization, technological development and innovation capacity, began to exert influence on regional development and reshape the regional development process and pattern in China. In addition, the way in which these new factors affect the formation and evolution of regional development pattern is different from that of the traditional ones, which renders many classical theories ineffective. For example, traditional spatial interaction theory is built on distance-decay law. However, the emergence of informatization and multinational companies expanded the scale of spatial diffusion of information, capital and technologies and reshaped the precondition of distance-decay theory. New factors formed underlying the formation of new industrial space, regional comparative advantage and regional competitiveness have restructured the patterns previously dominated by traditional factors, such as natural resources, transportation, geographical location and so on, while regional innovative ability, institutional advantage, cultural and value inclusiveness play an increasingly important role in regional development.

1.1.4 The Uncertainty of Regional Development — Effects of Human Activities

One of the distinctive characteristics of regional development is uncertainty owing to the significant impact of human activities on regional development. Human beings are important affecting factor of regional development, as the development philosophy, value orientation, psychological and behavioral patterns of human beings will have direct influence on the goal-setting, status and effectiveness, as well as path selection of regional development. In addition, inter-generational difference, inter-group variation and individual characteristics are expected to have varied effects on regional development, leading to its increasing complexity and spatial-temporal differentiation.

The influence of physical-geographical environment on the distribution of human activities was once the analytical focus in the history of regional development research. A case in point is environmental determinism which follows a passive view of human behavior. A parallel concept is "man will necessarily triumph over nature" which neglects the importance of natural law and overstates the role of human subjectivity. These two types of development philosophy were unfavorable to both regional development and regional development research. Nowadays, the emphasis on the harmonious relations between human and nature becomes the core concept of regional development. With the implementation of scientific outlook on development in China, human initiatives should be understood as an enabling factor in regional development.

As the first reflection of human initiative, goal orientation of regional development is determined by the development philosophy of human beings. Here "human" refers not only to humankind as a whole, but also to individuals and groups. From the perspective of humankind as a whole, development philosophy, measurement and other related theories of regional development will change in accordance with the shift of human value orientation. For example, economic growth used to be taken as the sole objective of regional development in the past. Such situation began to experience changes since the emergence of sustainable development concept, as social and ecological developments were incorporated into the objectives of regional development. Moreover, evaluation system, identification methods, and mechanism of regional development were changed accordingly to further affect regulatory behavior.

From the perspective of individuals, personal preference, psychological factor and behavioral habits that were once neglected in regional development research is now drawing increasing attention. The spatial difference in individual identification with social and cultural factors, such as institutions, values, ethics, customs, and ideology, contributes to the formation of territorially specific cultures in specific regions. These localized cultures have long-term impact upon people's behaviors, social relations and regional development. They are becoming a powerful driving force of regional development and playing an increasingly important role in socio-economic development.

Under such circumstance, the investigation of the process through which human subjectivity affect regional development has become a new direction of regional development research. On the one hand, basic theoretical research such as institutional turn and cultural turn in economic geography that reconstructs regional development theory based on the examination of human subjectivity can provide solid theoretical foundation for the scientific identification of human enabling forces underlying regional development process. On the other hand, human factors should be incorporated into traditional research that aims to meet national demand on regional development strategies in order to embody people-centered development philosophy.

In addition, one of the important roles played by human beings in regional development is their blindly intervention into many important geographical processes in general and human-geographical process in particular on the earth surface that blurred the evolutionary rules of geographical processes and contributed to unidentified pattern of regional development and disorderly process of regional spatial transformation. For example, regional decision-makers usually go against the basic theory of urban and urban system growth and tend to arbitrarily accelerate urbanization process, and take the expansion of urban area as the basic objective of urban development and planning. Such practice will inevitably lead to disorderly spatial and scale structures of urban system and bring about difficulty to urban development and regional development research[2]. In this sense, the examination of basic theories and driving mechanisms of geographical processes through the perspective of blind intervention of human being constitutes a point of major difference between regional development research and physical geography and other earth sciences.

1.2 Basic Perspectives on the Current Situation of Regional Development Research

The scientific and technological demands of regional development in China and the international communication with foreign scholars have greatly improved regional development research in our country in recent years. Significant progresses have been made including the development of theoretical system of territorial development centered on territorial function and spatial structure that contributed to the improvement of economic geography as a subdiscipline, increasingly publicity of regional development research through participating in and undertaking key consultation projects of regional planning and development strategy, gradually improvement of research capability in the transformation of regional development pattern through continuously involvement in research on regional sustainable development and man-earth areal system, increase of knowledge about new factors, mechanisms and types of regional development through investigation of new territorial space and other forms of industrial spatial organization, and the strengthening of modern technological support for data collection, spatial analysis, model application and outcome representation through the development and application of new technology and methods.In general, due to its cross-disciplinary attribute, regional development research pays more attention to natural resource and environmental foundation and the goal orientation of sustainable development, which requires not only the knowledge of spatial structure and territorial layout but also a strong capability in cartographic representation and geographic information system (GIS) application. Therefore, compared with scholars in other disciplines, economic geographers have great advantages in regional development research. The team structure and organizational pattern of the Chinese Academy of Sciences is very favorable to the assumption of key research programs of regional sustainable development through the organization of large-scale research groups. As a result, the research team of economic geography at the Chinese Academy of Sciences has become the academic center of regional development research in China.

Regional development has become a core social and economic issue for central and local policy makers. Academic researchers in our country have made great achievements in national strategic decision-making on regional development and regional sustainable development and the development of theories and methods through analysis of socio-economic development trends and their resource and environmental effects at all spatial scales as well as the dramatic changes in natural and socio-economic structures.

1.2.1 The Supporting Conditions of Functional Zones and the Indicator System for Functional Zoning

In 2004, the Academic Division of Chinese Academy of Sciences was invited by the National Development and Reform Commission of China to carry out several consulting projects. Based on long-term research on land development, regional development and the evolution of resources and environment in our country, the research group completed the consulting project of "National Functional Zoning and the Supporting Conditions for Further Development of Functional Zones in China". The research findings have played a key role in the making of the 11th Five-Year Plan of China. During the past years of plan implementation, National Development and Reform Commission has relied heavily on the participation of experts and researchers in regional development for the classification of four types of top-level major functional zones in our country. According to scientific principles and indicators, the functional zoning proposal of social-economic development and comprehensive governance in China in the next 15 to 20 years was produced based on systemic analysis of natural foundation, development goals at present stage and plenty of regional problems induced by long-term rapid growth in China. In addition, main functions and development principles of major functional zones and their supporting conditions were figured out. We as researchers are making out efforts to facilitate the implementation of the 11th Five-Year Plan of China[13].

1.2.2 Research on Regional Sustainable Development

Many research findings about resources and environment in natural science need to find an outlet in sustainable development. Therefore, the research direction of resource and environment has been expanded to include sustainable development. During the past years, Chinese researchers in regional sciences have done a series of research work with high visibility in regional sustainable development, including research on China's national conditions in the 1990s, research on the development strategies of agriculture and food in China, *Regional Development Report in China, Sustainable Development Report in China, China's Modernization Report, etc.* These reports and related work have generated tremendous impact on national development[14].

1.2.3 Theoretical Research on Spatial Complex, Cluster and Spatial Structure

We have conducted both theoretical formulation and case study of the evolution characteristics of China's new industrial clusters in general and new-tech industrial agglomeration in particular by tracing the development trend of rapid industrialization and urbanization. Economies and diseconomies of scale in spatial agglomeration have been illustrated through the examination of the structures, benefits and thresholds of cluster, leading to the formation of industrial clusters across the whole country [15]. Theory of spatial structure is the research frontier of economic geography and regional economics in the international academia. We elaborate the general law of social-economic development process from an imbalanced state to a comparatively balanced one on macro scale and highlight that spatial agglomeration is still the long-term development trend of our country based on the practical problems emergent at the present stage of natural foundation and societal development. In the development process of pole-axis-area, agglomerated area is still the advanced form of social and economic spatial transformation. Moreover, regional differentiation of natural structure is the other foundation of the coupling relation between social and economic spatial structures in China. On meso- and micro-levels, amounting research has been conducted on the changes of the affecting factors of land use and the spatial reconstruction of social and economic elements, which analyzes the main driving factors of spatial sprawl of metropolitan area and the spatial-temporal process of land use expansion and develops the indicator system and methods for the delimitation of peri-urban areas in China.

1.2.4 Theoretical Research on City-region Interaction and the Development of Extended Metropolitan Regions (EMR)

Supported by the projects of Development of Western China and the Revitalization of the Traditional Industrial Base in Northeast China, research in this area calculates the intensity of interaction among a series of large cities based on the theories of city-region and EMR development and analyzes their relative location, transport connection, supply chain, commodity flow, capital flow, and information flow (telecommunication flow), *etc*. Moreover, gravity model has been expanded to better delimitate and simulate regional development status. In addition, some exploratory research has been conducted on the criterion and indicator system of gateway city and the integration of urban economic region based on both theories and practices of EMR, which reveals that EMR is the most competitive form of spatial organization under globalization. An additional feature of EMR in China is that there exist close and vertical industrial linkages between the core region and the outlying hinterland[16].

1.2.5 Research on Database, Spatial Analytical Methods and Regional Simulation

Significant progress of scientific methods has been made in the application of GIS and remote sensing(RS), the establishment of database and graphic libraries for special purpose and the development of spatial analysis technique, when we attempt to meet national strategic demand and develop disciplinary theory. So far, the provincial-level database has been built up which covers 30 categories and nearly 200 indicators such as population, investment, comprehensive economy, agriculture, industry, *etc*., within a time span of past 20 years. During the research on regional development and urbanization, such database and graphic libraries are applied to analyze the interaction among cities and regions and the relationship between geographical pattern and process.

They are also applied to determine the attractive area of social and economic objects such as cities and ports, the optimal path of transportation, the spatial accessibility away from urban core represented by isograms and the intensity of spatial interactive forces of those objectives. Such analyses have equipped the research on man-earth areal system, coupling between natural and socio-economic factors and the comprehensive zoning of sustainable development with advanced technical methods. In terms of regional development simulation, some scholars have developed in recent years a regional simulator prototype of macro-economic policy, and have produced a simulating system of general equilibrium model of regional economy with independent intellectual property rights and a dynamic-agent simulation platform of man-earth relationship named GBASE.

1.3 The Return and Innovation in Mainstream Regional Development Research

1.3.1 The Return in Geography: from Regionality to Regional School

Ancient Chinese geography mainly took the form of local chronicles concerned with the records and descriptions of what a certain region has in its territory. Modern geography has further developed the research paradigm of regionality, placing equal emphasis on intra-regional structures, functions, interactions and the inter-regional differentiations and connections.

In the late 1950s, the research paradigm of regionality in geography was questioned by worldwide quantitative revolution[17]. Such revolution was quickly succeeded by newly emergent behaviorism, humanism, constructivism and modernism, because it overemphasized the role of mathematical model, underplayed the importance of human behavior in regional development, and broke away from the feature of regionality in geographical research. As a result, regional tradition returned to the core position of geographical research, partly because of the re-newed understanding of the nature and role of region, and partly because of the advent of a new era in western social sciences in the 1980s characterized by cross fertilization and mutual interaction. Region used to be regarded as the product of political and economic processes rather than the basic unit of social life such as market, country, family and enterprise, or the fundamental process of social dynamic. Since the early 1980s, it has been gradually accepted by mainstream social sciences that region should be treated as the basic unit of modern economy and social life. Meanwhile, other disciplines such as sociology, economics, politics, history, *etc.* have developed great interests in region during their theoretical interaction with geography, and began to take region as one of the most advanced organizational form for coordinating capitalist social and economic life and a crucial source of competitive advantage.

1.3.2 The Development of Mainstream Regional Research

After the Second World War, people's interests in regional research increased drastically in association with the emergence or ever deterioration of many regional problems. How to recover from the war damage and rehabilitate social environment for human settlements became major problems confronting governments all over the world. With the advent of globalization and the wide acceptance of sustainable development, regional development research began to pay attention to such issues as spatial organization, resources and environment, sustainable development, global change, *etc.*

Spatial planning was derived from urban planning and industrial and mining district planning in the 1920s and 1930s. As early as 1909, urban planning of Chicago in US had already contained regional factors such as regional transportation and open space, *etc.* The surge of urban planning in western countries was partly because of the substantial increase in automobiles holders and related process of suburbanization, and partly because of close attention to environmental problems. In the late 1940s and 1950s, spatial planning centered on spatial coordination of industrial and urban construction was carried out in some metropolitan areas and major industrial and mining districts, as a result of economic recovery and rehabilitation in many European countries[18]. Location theory, central place theory and growth pole theory were developed in many countries and spatial planning was gradually expanded to cover agricultural areas, tourism and leisure resorts, and economically underdeveloped regions. In the 21st century, spatial planning in western developed countries began to shift from physical planning towards social development planning with increasing attention paid to social, ecological and cultural factors. For instance, western European countries have made economic forecast on the territorial structure changes of economic development and the resultant environmental changes in order to fit spatial planning into the reality.

Resources and environment have greatly restricted the economic and social development of countries around the world and a series of theoretic problems in terms of regional sustainable development need to be investigated. Firstly, no agreement has been reached on the selection of relevant indicators and the determination of their relative weights to measure the coordinated development of population, resources, environment and economy. Secondly, how to measure regional development quality which involves the relationship between regional economic growth and social development, and the quantification of green gross domestic product (GDP)? The third question concerns sustainable development at micro-scale, as the differentiation of resource and environment can only be measured accurately at finer scale. Therefore, future research should build on GIS and RS methods to explore the possibility of carrying out accurate analysis by means of large-scale maps and high-resolution remote sensing images. The fourth question is related to the establishment of regional ecological and recycling economic system. In this regard, regional ecological system can be gradually constructed through the study of several issues including the establishment of economic structure from the perspective of ecology, the formation of inter-enterprise relationships and eco-industry park, green manufacturing in industrial production and green agriculture in agricultural production *etc*.

As the guiding principle of contemporary social progress, sustainable development reflects both the harmony between human and nature and the sense of inter-generational responsibility. At present, studies on sustainable development are mainly focused on the following aspects such as the resource compensation mechanism of sustainable development, urban sustainable development, sustainable development of agriculture eco-tourism and tourism resorts, measurement system of sustainable development, the impact of foreign direct investment (FDI) on sustainable development, and the ethics and values of sustainable development *etc*. Countries all over the world are faced with the same challenge of economic sustainable development, especially the imperative to promote economic growth and reduce poverty and employment pressure. Therefore, research focus should be placed on: 1) how to transform economic growth model? 2) how to establish a harmonious society and reduce inter-regional and urban-rural gaps? 3) how to improve scientific and technological innovative ability and provide support for sustainable development?

Since the 1990s, research on global changes has entered a new period with significant adjustments of research directions in three aspects from early prevention and alleviation to later adaptation. The first aspect is the shift of research questions from the basic theories of earth system to a series of practical problems closely related to the sustainable development of human society. The second aspect concerns the extension of research focus from human impact on environmental changes to human adaptation to global environmental changes. The third aspect is to conduct integrated research on earth system at higher level.

1.3.3 The Formation and Resurgence of Regional Sciences

Regional Sciences came into being in the 1950s as featured by the founding of "Regional Science Association" in the United States. The objectives of regional sciences are to follow scientific methods and provide reliable theoretical foundation for regional policy-making, regional planning, regional development and other forms of regional analyses[17].

The development of regional sciences has gone through three stages. Before the 1960s, traditional regional sciences mainly described regional activities and established intra-regional and inter-regional models with the application of general equilibrium analytical method. In the 1960s, the focus of regional sciences was changed to the development of individually or generally regional decision-making models, which entails the analysis of appropriate regional income, employment, and investment, *etc.* by means of static or dynamic and linear or non-linear programming techniques. The third stage of "new regional sciences" started from the 1970s with a shift of research direction towards regional planning and policies. Theories on regional planning, regional economics, public policy, consulting and decision-making have become the core research area of regional sciences in recent years.

Regional Sciences have the following characteristics. Firstly, they emphasize the deduction from the general to the specific, attaching importance to mathematical analysis and theoretical modeling and simulation. Secondly, they advocate integrated and multi-disciplinary research. Thirdly, they emphasize the research on regional development policies and countermeasures to ensure the achievement of regional development goals.

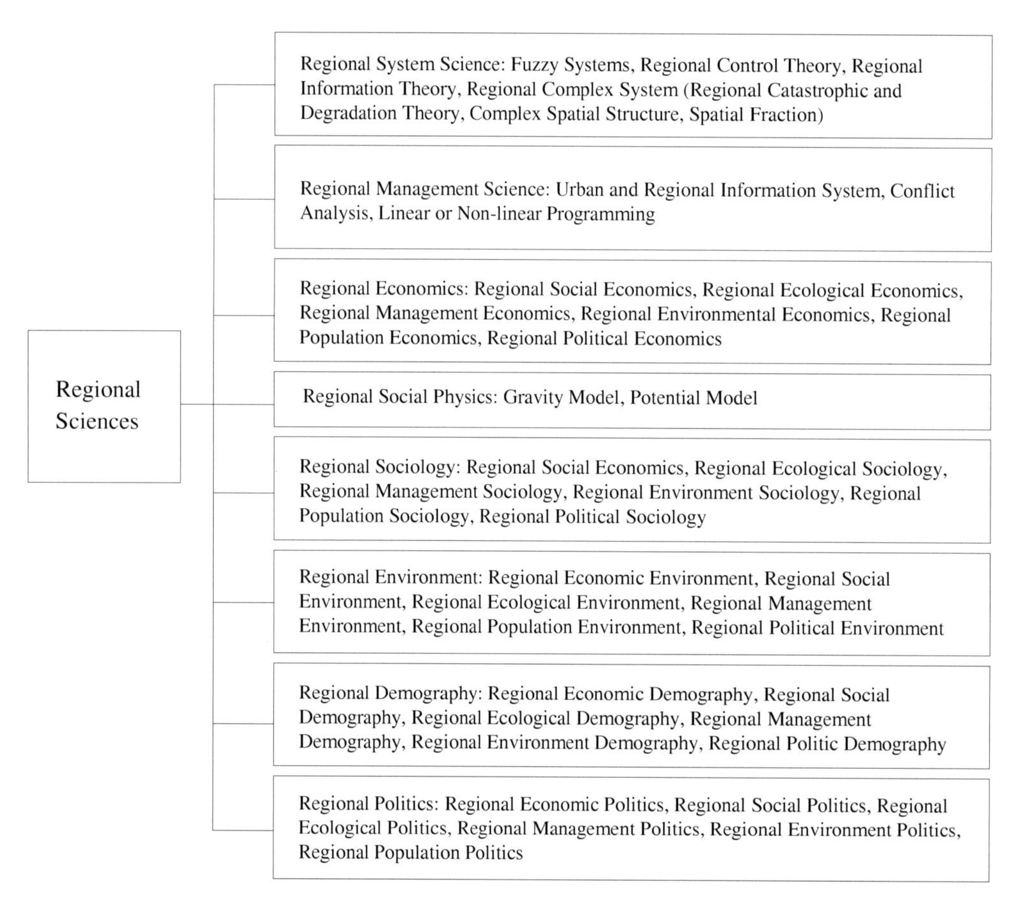

Fig. 1.2 The Structure of Regional Sciences

2 Basic Trends and Scientific and Technological Demands of Regional Development

The scientific and technological development in the field of regional development is largely determined by national and social development stage and the evolutionary trend of natural structure and features. When we analyze the mission, important scientific problems, disciplinary direction, methodology and suggestions of regional development research, it is essential to predict the main trend of future national development and its differentiation at different stages[12,19]. A series of important issues can be drawn from the following basic trends.

2.1 Basic Trends of Regional Development

2.1.1 Future Economic Development in China

During the additional years to 2020, economic growth in China will still maintain high speed[20,21]. It will step into the period of stable growth around 2020 and enter the slow growth period after 2030. The national economic aggregate of our country may reach 120–150 trillion yuan in 2050. The third-step strategic goal in China is to achieve the level of moderately developed countries and accomplish modernization by and large around 2050. The GDP per capita of our country in 2020, 2030 and 2050 will be 33 thousand yuan (6500 US dollars per capita), 69 thousand yuan and 123 thousand yuan respectively by constant price of year 2000. In 2050, The GDP per capita will reach the present level of developed countries, which approximately equals two thirds of per capita GDP of those countries at that time.

In the past 30 years, the rapid economic growth of our country can be attributed to the following four major factors: expanding scale of capital investment (including foreign investment); dramatically improving productivity brought about by institutional reform and technological progress; the "infinite" supply of cheap labor (*i.e.* "demographic dividend"); and significant increase of imports and exports through participation in economic globalization (*i.e.* "world factory"). In the next 50 years, the capital input of our country can still remain at a high level against the background of global surplus of capital, while the other three factors, namely institutional reform effects, "demographic dividend" and "world factory" will go through great changes. In this case, economic growth of China will slow down and enter the stable growth period gradually[22].

Firstly, after nearly 30 years of reform, the role of market mechanism has been brought into full play, and the efficiency released by further reform will be decreasing. Secondly, the "demographic dividend" will come to an end. The number of newly-added working-age population decreases year by year and will reach zero growth rate around 2015. The rise in labor cost is inevitable, and the share of aging population and dependency ratio will increase rapidly. Thirdly, the contribution of "world factory" growth model to Chinese economic growth will gradually decrease due to the influence of resource and environment, RMB appreciation and the disappearance of "demographic dividend". It is predicted that Chinese exports will step into the slow growth stage after 2015 (with the peak value of external dependence degree at about 100%).

In sum, it is predicted that the average annual growth rate of Chinese GDP in 2008–2020, 2020–2030 and 2030–2050 will be kept at 7%–8%, 5% and 3%–4% respectively, and by 2020 the Chinese GDP aggregate will be up to 47 trillion yuan (at constant prices of year 2000) which is 2.5 times as much as that of 2007, if not encountered any global economic crisis. In consideration of RMB appreciation, China's economic aggregate will rank No. 1 in the world (measured in US dollars) with a value of nearly 10 trillion US dollars in 2020. The GDP aggregate will reach 98 trillion yuan in 2030 and 152 trillion yuan in 2050, which are 5.2 times and 8.0 times as much as that in 2007 respectively.

According to the differences in the growth rate of three industries in the past and their future trend of change it can be estimated that the income elasticity of demand of the secondary industry will continue to increase in the next 10 years[23,24]. Its proportion in the national economy will decrease gradually after 2020, but the pace of decrease will be slow before the saturation of demand for per capita material products. It is predicted that the structure of three industries will change as follows:

- in 2020, 11 ∶ 50 ∶ 39
- in 2030, 9 ∶ 48 ∶ 43
- in 2050, 6 ∶ 42 ∶ 52

It is estimated that after 2020, the growth speed of material consumption and the consumption of traditional service will tend to slow down, while experience-based consumption (*e.g.* culture, film and television, tourism, leisure, vacation, endowment, environmental consumption, network experience consumption, *etc.*) will become the main source of consumption growth. The transformation of consumption structure will lead to changes in regional competitive advantages. Certain areas with no advantages in traditional manufacturing industry may become new economic growth poles (such as the

"Sun Belt" in the US), while some areas with manufacturing advantages will probably turn to develop new industries (such as Los Angeles in the US), or become problematic regions (such as Detroit in the US).

2.1.2 Chinese Population Growth, Population Flow and Urbanization Process

Given no change of birth-control policy, it is predicted that the total population of our country in 2020 will be 1.43 billion, and will reach its peak value at 1.45 billion around 2030. After that, it will go down to 1.38 billion in 2050.

Over the past decade, China has entered an immigration period (immigration period refers to a period in which the proportion of migration population in total population keeps steady over 10%)[25,26]. Between 1990 and 2000, the average annual growth rate of floating population in China was 15.5% and the number of floating population had increased by 110.26 million during the decade. In addition, according to the latest data of 1% population sampling survey in 2005, Chinese floating population reached 147.35 million in 2005, an increase by 2.96 million compared with that of year 2000, and inter-provincial floating population accounted for 29.4% of total floating population. At that time, urbanization process in China reached its climax. During the process of inter-provincial migration, provinces in Yangtze River Delta such as Shanghai, Jiangsu and Zhejiang, and developed areas along eastern coast such as Guangdong and Beijing become the major destination of floating population. The immigrant population mainly comes from central and western provinces and cities in close proximity to these open regions. In the next 10–20 years, the general pattern of population flow at national level will not change but the scale of population flow will decrease. At meso-level, the scale of population flow from rural to urban areas within each province will increase and vary greatly from province to province.

In view of an annual increase of urbanization level by 0.8 percentage point, urban population will be 0.79 billion (with urbanization level at about 55%) in 2020. After that, supposing that an annual increase of urbanization level by 0.7 percentage point, urban population will reach 0.95 billion in 2030 (with urbanization level at about 65%, and urbanization process will slow down afterwards); supposing an annual increase of urbanization level by 0.5 percentage point after 2030, urban population will reach 1 billion in 2050 (with urbanization level about 73%). Therefore, urban population and urbanization level in China may reach their peak values in 2050 when urban population will be around 1 billion. In the following 40 years, nearly 0.4 billion rural population will be transformed into urban population[3,27].

Metropolitan regions and population and industrial agglomeration areas are in the process of formation in China and increasing problems emerged in terms of their optimal organization and coordination with natural and environment foundation.

2.1.3 The Severe Situation Confronting the Evolution of Resources, Environment and Ecology in China

China will be faced with increasingly severe resource shortage and environmental crisis in the next 20 years. The ecological and environmental conditions might be under protection and control after 2030. Under ideal conditions, resource-saving society will be formed by and large and environmental and ecological pressure will be relieved considerably in 2050. Such development implies that regional development and social-economic growth in China will face more constraints in energy, resources, environment, and ecological conditions in the long run. In other words, we are still confronted with double pressure from promoting economic growth and protecting ecological environment.

It is extremely difficult to reverse the trend of environmental deterioration in the next 20 years[6,29]. The reasons are as follows: Firstly, efforts must be made in our country to maintain a moderate growth rate for a long time. Secondly, the middle stage of industrialization will last for a certain period. At this stage, industries with high intensity of resource consumption and large amount of waste emission will continue their leading position. Thirdly, Chinese population will continue to increase in huge amount and large-scale urbanization will reshape consumption structure, consume additional energy and water resources and generate additional amount of waste. Fourthly, it is impossible to spend a large amount of money on environmental treatment and ecological construction, as per capita income in our country is still low. However, the government and the society should implement proper strategies to ameliorate the pressure of environmental crisis. After the 2030s, there will likely be real improvement in ecological and environmental conditions in China, leading to the coordination between economic growth and ecological environment.

It is quite possible for China to maintain a rapid economic growth rate in the next 20 years. And the whole economy will continue to consume huge amount of resources without a significant improvement of technological level[30,31]. Resource problems are closely related to ecological environment. The trend of environmental deterioration will be reversed as national economy turned from high-speed growth to moderate and steady growth. In other words, after 2030 when China steps into the late stage of industrialization, the level of social welfare will be improved considerably, the environmental deterioration tendency will start to be reversed, and economic growth will stabilize at normal rate.

According to national conditions and the general trend of social and economic development in China, it is necessary to gradually establish a social-economic system with full utilization of human resources and frugal use of natural resources in the next 20 years. We need to improve infrastructure supply and ecological and environmental conditions as soon as possible, establish an industrial and technological system with late-comer advantage in certain exports, follow the urbanization model with Chinese characteristics of huge amount of rural population, accelerate the development of central and western areas while promoting the modernization of coastal areas and traditional industrial bases in order to alleviate enlarging regional gap. This development strategy, which is centered on enhancing the quality and benefit of economic growth, can help to relieve the eco-environmental pressure imposed by economic and social development to a large extent.

2.1.4 The Development of Infrastructure System (Transport and Information)

Over the past three decades, the advancement of transport and information technology and the construction of basic infrastructure have provided solid support for regional economic and social development in China. They played an important role in the restructuring of regional development pattern by guiding rational economic flow and optimal distribution of cities and towns, optimizing regional industrial structure and promoting industrial transfer. Transport system becomes a key factor for establishing orderly structured regional development pattern, optimizing the sequence of land development, and cultivating hub cities or gateway cities. It helps to enhance the international competitiveness of Yangtze River Delta, Beijing-Tianjin-Heibei area, and Pearl River Delta, and uplift the locational advantage and global position of hub cities such as Beijing, Shanghai, Guangzhou, Shenzhen, Tianjin, *etc.*[32,33]

The next 20 years will be an important period for further development of transport and communication technology. Key directions of development include large-size and high-speed instruments, orderly and universally applicable network, convenient services, and timely and efficient management system. Judged from its supporting conditions and development trend, the planned network skeleton of high-speed railways and expressways will be formed by and large in 2020, and aviation and information networks will catch up with advanced world levels. The transport and information technology of our country will occupy first place in the world, and advanced transport system and information network will also be formed in 2050.

From 2020 to 2030, comprehensive transportation modes with large capacity will be constructed among major urban agglomerations in China. High-speed railways, intercity rapid railways, magnetic levitated railways, *etc.* with a top speed of more than 500 km per hour will continue to be built. At the same time, the communication technology will be further modernized, which forms rapid inter-city connection networks. The utilization of super-large transport ships, large-capacity communication equipments and airplanes will further facilitate the connection between our country and the rest of the world.

The advancement of transport and communication technology will further intensify two forces affecting regional development in our country, namely the agglomerative force and dispersion force. Firstly, e-business, virtual economy, logistics economy, *etc.* will become important driving forces of regional development. Secondly, effective, intensive and sustainable development of territorial space will become the basic developmental model for specific regions which emphasizes rational, economical and intensive use of land resources (*e.g.* cultivated lands, coastlines, water areas, and airspace areas) through resource integration. Thirdly, further growth of hub cities and gateway cities will be achieved. Fourthly, the driving mechanisms at intra-regional level will experience change. All these trends will complicate the mechanisms of regional development.

Informatization will bring about more profound changes to social economy. Most importantly, it will lead to the spatial restructuring of economic and social forms. It developed firstly in a few developed countries in the 1970s and 1980s and has exhibited high-speed development in China since the 1990s [34].

Information has become increasingly an important productive and location factor and time costs become more and more significant. Informatization has promoted the diffusion, application and innovation of knowledge. Informational economy will continue to strengthen the existing pattern of spatial polarization in economy and society. The development of new phenomena of regional differentiation such as digital divide, digital differentiation, *etc.* has enlarged spatial disparities at macro level. There exist significant differences in informatization, especially the differences in ability to master, spread, and utilize information among three macro regions in China, which have conspicuous impact on regional variations of economic growth.

2.1.5 Economic Globalization Will Be a Tortuous and Gradually Deepening Process

Since the reform and opening up, China has made full use of the developmental opportunities brought about by economic globalization to achieve continuous and rapid growth of national economy and significant enhancement of comprehensive national strength. Meanwhile, the overwhelming trend of economic globalization is also strengthening the imbalance of economic development among different regions in China as a result of the competitive force of "survival of the fittest". Economic globalization exposes regions to direct global competition and the region is not only affected by domestic economic conditions, but also by global economic changes[10,35]. According to a statistical analysis, the contribution rate of import, export and foreign investment to the inter-provincial GDP differences in China is over 20%. These factors have strong correlation with economic location and are very difficult to change.

During the next few years, low production cost and large market scale will still constitute the competitive advantage of our country to attract foreign direct investment. Moreover, low production cost and large economies of scale will contribute to the competitive advantage of international trade in China in the near future. But in the long run, it will be very difficult for foreign trade to keep a growth rate surpassing GDP by a large margin. On the whole, the degree of China's dependence on foreign trade will reach the peak and then slowly go down around 2015.

In the future, foreign investments will still mainly flow into coastal areas and those focused on tapping into Chinese market will probably flow to inland areas along major comprehensive transport routes. In general, it is more likely that foreign investment will flow primarily to coastal areas, areas along the Yangtze River and then to other places, the spatial pattern of which appears to be a "T-type". Influenced by the maritime transport conditions, the spatial pattern of foreign trade is basically identical with that of foreign investment. Therefore, the impact of economic globalization on regional development in China is to consolidate the "T-type" spatial pattern and to facilitate the development of a few metropolitan regions with international competitiveness.

In addition, owing to the continuous enhancement of the economic strength of our country, the overseas investments of China will increase rapidly. According to the growth experiences of developed countries, China's surrounding countries (especially Central and Southeast Asian countries) are likely to become key destinations of oversea investments. Special economic zones similar to the border manufacturing regions between USA and Mexico will possibly be formed in adjacent areas of China or some new economic centers may emerge in the border areas at least.

Under the development of economic globalization and informatization, the formation of metropolitan economic regions (global cities, national cities, and regional gateway cities) leads to new imbalance of development. The "space of place" of world economy is being replaced by the "space of flow". The current spatial structure of world system is built on the logical framework of "flows", links, networks, and nodes (management framework is transformed from vertical, pyramid-type to flat, flow type), which results in the high density and spatial compression of economic activities. One important consequence is the formation of "gateway city", which serves as the confluent point of various "flows", the node connecting region with the rest of the world, and the control center of economic system. Under the background of economic globalization, "city-region" composed of "gateway city" and its closely connected hinterland is becoming the basic unit of regional economic competition around the world.

Such territory units with global or regional competitiveness are also in the process of formation in China. The prospective research and planning on their location, industrial structure, spatial structure, and supporting system will be a long-term research work for us.

2.1.6 Regional Development in China under Global Change

In order to avoid the irreversible deterioration of global climate, the global greenhouse gas emission reduction is becoming a worldwide consensus, which brings both opportunities and challenges to the adjustment and upgrading of regional industrial structure[10,35,36].

Under the pressure of carbon emission reduction, China will gradually establish green trade system, restrict the export of high energy-consuming and high carbon-emission products and industries such as rubber products, mining and metallurgy products in the future, and encourage the growth of "energy-efficient, environment-friendly" industries such as high-tech products. Such action will definitely exert influences on the spatial distribution of the foreign trade in China.

Low carbon economy, based on low energy consumption and low pollution, is considered to be the new growth engine in the future and the fifth revolutionary force of the world economy following the industrial revolution and the information revolution. New energy, clean energy, biological energy, renewable energy, and so on are all part of the low carbon economy. Against the background of global greenhouse gas emission reduction, the amount of carbon emitted will become an important affecting factor of regional development, and regional economy with low-carbon industrial structure has strong potential to achieve sustainable development in the future.

Of course, there are still some uncertain factors and processes underlying the above trend that defy imagination and prediction during the next 50 years. For example, international monetary system may be changed. The current international monetary system is a kind of floating system centered on US dollar and diversified currency reserves. The transformation of such system in terms of the restructuring of global financial system and the spatial transfer of global financial centers will bring new opportunities to China's regional development, as one or two global city-regions or international cities based on advanced business services such as finance and commercial trade are likely to take shape in China. We should make initial efforts to make planning and other necessary preparations.

2.2 Scientific and Technological Development Tasks in Regional Development

Key tasks confronting regional development in China include not only analyzing affecting factors and developing trend, but also providing suggestions and policy recommendations. In view of the imperative of inter-country and inter-regional competition and cooperation and the need to improve the rationality of decision-making, regional development research will have to shift the research focus from tracking developing trend to forecast and prediction in order to better serve scientific decision-making in regional development and achieve coordinated development of man-earth relationship. Here the "development trend" covers broadly to include economic growth, social development, structural changes, investment effects, and the evolution of ecological environment and so on.

According to the above-mentioned prediction, our country will become a large economy in the world with unique features. What are the composing structure of this huge economic entity and its internal spatial structure? The answering of this question will exert significant influence on the spatial pattern of regional development in China. Undoubtedly, the development of such a huge economy will have to solve a large number of regional development problems. For instance, how to maintain the competitiveness of core areas? How to sustain economic growth with the support of resources, environment and basic infrastructure? How to solve the long-existing problem of imbalanced regional development?

In the next 15 to 20 years, Chinese economy will remain at a high development level, and central and regional natural and socio-economic structures will continue to evolve. The coupling and adaptive relationships between social economy and natural basis may become degrade and compel us to confront with severer problems of territorial space security and resource guarantee. How to maintain the survival conditions of ecologically fragile and environmentally deteriorated areas? How to improve the sustainable development capability of metropolitan regions and population and industrial agglomerated areas? All these problems are closely related to territorial space and resource security and long-term survival of our country and nation.

According to the current development trend, the aggregate of our national economy will reach an enormous scale in the next 40 years. However, it is definitely not an easy path of growth and many difficulties need to overcome, among which regional development imbalance and various types of resource and environmental problems confronting different regions stand out. Based on national economy and social development demands mentioned above, regional development in China is confronted with the following major tasks in science and technology development:

(1) Monitoring the dynamic status of regional development and the situation of regional imbalance, building up an early warning system for regional development, and controlling regional imbalance to avoid social unrest.

(2) Monitoring the ever-shifting resource-environmental problems confronting regional development, especially the developing state of major function oriented zones to provide scientific support for decision-making on regional development.

(3) Tracking new affecting factors of regional development and providing guidance on economic adjustment and spatial reorganization of different types of regions in order to maintain long-time competitiveness.

(4) Studying urbanization issues in regional development, directing the reasonable and orderly flow of population, and providing decision-making support for scientific and stable urbanization process.

(5) Strengthening theoretical and policy research on "problem regions". In the next 40 years, with the changes of development stages and macro economic situation, "regional reshuffling" is an inevitable phenomenon on the road to modernization. The redevelopment of "problem regions" is a global topic in

need of further exploration.

It is possible that the forecast and prediction of regional development will become true in the next 20 years, in conformity with national and regional decision-making needs. It is expected that regional development research should be able to reflect the demand of harmonious development of man-earth areal system and to more accurately measure the process and situation of regional development in 2050.

2.3 Scientific and Technological Demands of Regional Development in China (to 2030)

2.3.1 Spatial Pattern of Economic Growth and Social Development at Major Developmental Stages

Our country has been confronted with the threat of land resource security. Given the present situation and developing trend of social economy, serious problem of land resource insecurity will emerge. Spatial agglomeration will very likely to become the dominant tendency before 2030[1,19,37–39].

What is the general strategy for future regional spatial development in China? Which regions will probably be the agglomeration areas of population, cities and industries? In which regions should protection and consolidation policies be adopted? How to achieve the general strategy of regional development? The ongoing project of functional zoning conducted by relevant state departments, of which regional development research scholars are a part, is the beginning to solve all these problems[13,19].

The present trend of spatial distribution of socio-economic development cannot sustain. Many special regions such as the vast area of Tibet Plateau, the Loess Plateau, northwest arid and semi-arid areas, farming-pastoral transitional zones and karst areas, *etc.* in our country are ecologically fragile areas. Some of these regions are severely short of water and land resources. Therefore, in these regions, large-scale industrialization and urbanization should be avoided[13] and high GDP growth target cannot be achieved for long time. As a result of spatial agglomeration, competitiveness enhancement and internationalization, the extent of industrial and population agglomeration in large cities and metropolitan regions will further increase. In this case, the resource supply conditions and eco-environmental carrying capacity of large and medium-sized cities and metropolitan regions need to be improved significantly.

Under such circumstances, the key areas for future economic growth and continuous urbanization must be those with favorable conditions of climate, water and land resources. Such places mainly include coastal areas, central and western plains and basins. Large-scale industries and urban population have concentrated in Pearl River Delta, Yangtze River Delta, Beijing-Tianjin region, central and southern Liaoning province, *etc.* In view of international experiences, the ongoing processes of internationalization and informationization will further expose man-earch system in these areas to the external world and such areas have great potentials to become territorial space characterized by high-density, high-efficiency, resource-saving and modernization, under the support of modern infrastructure[3,40].

Disorderly territorial development and uncontrolled urban sprawl should not continue. Our country has a vast territory yet with inadequate resource per capita. If the current situation of land destruction, water pollution and large-scale consumption of water, energy and mineral resources in economic and social development cannot be improved, there will be no land to occupy in some municipalities in the next ten years. Per capita cultivated land in some densely populated and economically developed provinces will decrease to below 0.3 *mu*①, and the self-sufficiency rate of grain in developed coastal areas where more than half of the total population of our country inhabit will drop to below 70%. This will pose serious challenge to the sustainable development of future generations. If the limited space and resources cannot be utilized rationally and urban sprawl cannot be curbed, the processes for the construction of well-off society and national modernization will be considerably delayed[3].

Issues on the imbalance of regional development

The differences in the level and structure of regional development will be a long-lasting trend in our country. The Government has to make continuous choices between balanced and imbalanced development. Scholars should simulate the actual process and provide theoretical explanations. How should we understand regional differences in terms of economic development level and strength?

The affecting or even deciding factors of spatial disparity in China include natural foundation and natural resources, historical legacy, natural or historical location, science and technology innovation and institution reform. The development of globalization and informationization in recent years are the main factors that contribute to the widening of the gap. Major forces underlying the formation and evolving trend of regional disparity in our country include:

- Significant differences in the natural basis of different regions in China. Some key factors cannot be changed by human efforts.
- The present development stage in China: High-speed economic growth and relatively low economic aggregate per capita.
- Globalization and informationization have exacerbated the development imbalance of different regions.

① 1 *mu*≈666.7m^2

• Theory and practice: It is a long process to change from imbalanced to comparatively balanced development.

Most of the resources per capita in China are lower than the world average . Twenty years of rapid economic growth after the reform and opening-up has consumed or occupied a large number of natural resources, which put our country under tremendous resource pressure. Future economic growth will be confronted with severer threat of resource shortage, and expanding resource import will also be confronted with increasingly serious international political pressure and conflicts. Therefore, China needs to pay great attention to the saving and intensive use of resources during regional development and to build up a resource-saving socio-economic system with relatively concentrated and high-density spatial structure.

In response to the above problems and serious situation, we have to reorganize the regional space of our country gradually, make planning on the pattern of territorial space exploitation and spatial distribution, minimize man-earth contradictions emergent in the institutional transition from plan to market, coordinate the constructions of key projects in close association with China's long-term development, and rationally make use of natural resources and ecological assets in our country in order to avoid the emergence of severe situations of territorial space security that may endanger the survival and development of our nation. Of course, this is a long-term and step-by-step process of research and implementation.

Regional governance should be strengthened in order to implement the Scientific Outlook on Development. Territory is an organism with certain functions. Different territories have different features and functions. Rapid economic growth and large-scale urbanization of our country requires customized control or governance of territory, which is in turn based on various types of regionalization such as natural regionalization, ecological regionalization, economic regionalization and functional zoning, *etc.* [13,19,41–44]. Some strategic questions need to be investigated for the exploitation, conservation and development of territorial space in China, namely:

China's economic development will remain at a high level in the next 15 to 20 years, and both central and regional natural and socio-economic structures will continue to evolve. The coupling and adaptive relationships between social economy and natural basis may degrade, which may compel us to confront with severer problems in territorial space security and resource guarantee. How to maintain the survival conditions of ecologically fragile and environmentally deteriorated areas? How to improve the sustainable development capability of metropolitan regions and agglomerated areas of population and industries? All these problems are closely related to territorial space and resource security and long-term survival of our country and nation.

In the next 10 to 20 years, the spatial structure of Chinese socio-economic development will show a tendency toward spatial agglomeration. The extent and process of agglomeration need to be investigated further. In general, the formation of metropolitan regions and population and industrial agglomeration areas will occupy a center stage in the research and prediction of national development trend.

Main trend of territorial development and economic planning will be relative equilibrium after 2030. However, it is worth making prospective predictions on the development level and form of such equilibrium.

2.3.2 Rational Progress and Mode of Urbanization in China

Urbanization is a significant issue of national and regional development. It is imperative to study this issue from broad and multi-disciplinary perspectives. Over the past 10 years, China's urbanization process exhibited the feature of rash advance and disorderly structure. Important resources such as cultivated land and clean water have been excessively consumed, and the environment has been severely polluted which has brought about increasingly negative effects to the sustainable development of our country [3].

It is necessary to understand the national conditions of our country in order to address the core issue of urbanization. China is a country with large population and rapid expansion of economic aggregate, yet a serious shortage of land and water resources. China's urbanization process should not follow the pattern of large-scale urban sprawl in the United States. It is necessary to analyze and evaluate the driving factors of urbanization and urban development, the speed of urbanization process, urban-rural relationship, and the resource and environmental basis of urbanization process in those middle-industrialized, late-industrialized or post-industrialized countries from both national and regional perspectives in order to achieve a better understanding of their growth experiences. Any urban development planning or urban system planning from the singular perspective of urban development may lead to unimaginable outcome without an appreciation of national conditions and regional characteristics and differences. In this regard, we need to answer the following three practical and theoretical questions:

(1) Whether economic growth and the increase of employment positions can sustain China's high urbanization rate in the future? What influences will the poor carrying capacity of resources and environment have on urbanization process? What kind of urbanization model will be compatible with China's industries and carrying capacity of resources and environment?

(2) What is the indicator system and management method of resource-saving urbanization model?

(3) Many countries have undergone a long process of urbanization. Based on our national conditions, total population and industrial base, is it possible for urbanization process in our country to outpace that in developed countries?

The process and pattern of China's urbanization will be a Chinese style rather than the copy of European-American type. However, the speed of China's urbanization cannot drift away from the principle of gradual improvement. Future urbanization in China must follow resource-saving development model featured by high-density, high-efficiency, resource-saving and modernization instead of the mode of large-scale urban sprawl in the United States. Therefore, future work needs to be done to clarify the development trajectory of such mode. China can never follow the examples of western developed countries in terms of resource occupation and per capita resource consumption and academic scholars should put forward index system with Chinese characteristics to measure the scale structure, spatial structure and resource occupation amidst China's urbanization process in response to the phenomena of large-scale planning and urban sprawl at the present stage.

Urbanization entails a fundamental change of the economic structure, social structure, mode of production, and lifestyle of a country. Therefore it is necessarily a gradual process of accumulation and development. What is the gradual development of urbanization? How to enhance urbanization level on the basis of urban industrial absorptive capability, the supporting capacity of urban infrastructure, the carrying capacity of environment and resources, and the degree of the improvement of urban management level? China should not follow western countries in terms of resource occupation in socio-economic development. Instead, we should always live a thrifty life even at the stage of high-level modernization.

What is the reasonable urbanization rate of China in the future? To achieve the aims of urban-rural coordinated development and common prosperity, it is important to expand urban employment to absorb rural surplus labor force and allow more rural population to enjoy urban civilization through the increase of urbanization level on the one hand, and to narrow urban-rural gaps on the other hand through developing rural economy, constructing rural society, improving rural production and living conditions, and enhancing farmers' educational level. The optimal choice for China is to combine healthy urbanization process with new rural construction. In view of China's basic conditions of huge rural population, conspicuous contradictions between urbanization and cultivated land protection, tremendous urban employment pressure and near-saturated carrying capacity of environment and resources, the urbanization rate in China does not have to achieve the standard of developed countries, *i.e.*70% to 80% or even higher. The optimal level of urbanization needs to be further discussed and investigated.

China's urbanization should strictly follow the principle of gradual improvement, and adopt the mode of resource-saving development. We should conduct long-term research on the process, evaluation system, scale structure, and regional differentiation of urbanization in China in order to substantiate such development path.

2.3.3 Mechanisms and Trends of Regional Development in "Three-Dimensional Objective Space"

In order to achieve more coordinated national and regional development, the philosophy, objectives, processes and governance policies of regional development all call for adjustment and change. Both government and academia should incorporate resource and environment protection into the objective system of regional development in face of the enormous costs and pressure of resources and environment. Meanwhile, population growth, education, medical care, population flow, aging population, urbanization, fair distribution of personal income and so on are gradually becoming the main objectives of regional development. In this case, we should redesign the philosophy, objectives and policies of China's regional development and provide scientific support for central and local decision-makings over regional development strategies and policies through the revelation of regional development direction and inter-objective and cost-benefit relationships. The following scientific questions are to be investigated:

(1) Interaction among economic growth, social security and eco-environmental protection in "Three-Dimensional Objective Space (3DOS) ".

(2) The characteristics and affecting factors of economic growth curve in "3DOS". The direction and process of regional economic growth in "3DOS" are different from those under single economic goal.

(3) The process, structure, effects and governance of regional development in "3DOS". Such research entails an exploration on the basis of international experiences and China's development practices.

In particular, the revelation of the interactive relationship among three targets in "3DOS" and the evaluation of the comprehensive effects of regional development are the core issue of this research. We should establish the analytical framework of affecting factors, directions, processes and comprehensive effects of regional development in "3DOS", illustrate that "3DOS" is the advanced form of regional development and predict the general development trend of various regions in China under the framework of "3DOS".

Regional development in the "3DOS" pays equal attention to three development targets, namely, economic growth, social security, eco-environmental protection. As a result, it entails "smart growth" of our country and regions. In particular, social development involves not only social security, but also individual satisfaction and personal development. During this process, the pressure of eco-environment will experience an evolving trend of initial increase and later gradual decrease. Therefore, we need to develop the theory of regional economic growth and the methods of simulation and analysis under this objective system.

2.3.4 Application of Simulation Technique and Spatial Analysis

Simulation technique and spatial analysis are widely applied, but the analytical models are mainly applied at global or national levels. Analytical models of sustainable development status at regional level are still lacking without clear illustration of the characteristics of regional sustainable development. As a result, the analysis and model building of the status of regional development are too general to provide decision support for regional planners. Therefore, research on the simulation analysis models of regional development status in correspondence with China's development stage constitutes the research frontier of regional development and urbanization.

Internationally, analysis of economic system based on computable general equilibrium (CGE) is a hot topic in the simulation analysis of regional development. CGE model is mainly applied to the analysis of the influences of policy and economic changes at a national level. However, with the development of CGE research, how to apply the CGE model to a regional level has become a prominent problem. One important area of regional simulation is urban evolution which requires multi-disciplinary knowledge of geography, economics, complexity science and computer science and so on.

It is believed that in the next 20 years, research based on GIS, CGE and autonomous agent simulation in combination with other models will experience widely development in regional simulation. Meanwhile, simulative analysis based on the general dynamic models that has been widely used by Chinese and foreign researchers for a long time may make further progress. Such models are mainly applied to decision-making analysis and policy simulation.

In the next 50 years, a breakthrough in computer intelligent function may be achieved. Therefore, there may emerge three tendencies in regional development simulation. Firstly, simulation of the interaction among multiple regions will be carried out, and the concept of regional simulation from a global perspective will be formed under global economic integration. Secondly, theoretical research on regional development will be developed, some of which will be supported by computer simulation. In this sense, computer will become a tool of theoretical inference and trend forecasting and simulation technique at that time will definitely go beyond the framework of GIS-CGE-ABS. Thirdly, the combination of simulation technique with remote sensing or other new techniques will become a tool for instant analysis of regional environmental and population management.

The main objective of experimental analysis of regional development is to make forecast of the evolving trend of ecological environment as a result of territorial and regional development. It is possible to present to both government and the public on yearly or quarterly basis the trends and problems of regional development by means of "man-machine conversation". In the next 20 years, the core task of sustainable development research center at CAS is to build "Regional Development Laboratory".

2.3.5 Regional Problems and Development from Global Perspective

We should adapt to the new development trend of economic globalization and institutional reform in our country, further develop global perspective, actively participate in international economic and technological cooperation and competition, and improve the comprehensive level of opening-up. President Jintao Hu emphasized that we should develop global strategic vision[10,37,40]. In the future, globalization will continue to exert great influence on economic growth in China. The Population and economic activities will increasingly be concentrated in coastal area and the role of economic centers and metropolitan regions will become more prominent.

To secure an adequate supply of energy and other important resources, it is important to strengthen the cooperation between China and other countries in general and the surrounding countries in particular, and to promote the development of international cooperation regions at different spatial scales, which will become a key issue in the areas of economy, resource and environment and geopolitics.

2.3.6 Ecological Compensation and Financial Transfer Payment

We should be highly involved in national and regional governance. The implementation of regional policy of equal access to public services and scientific regional development strategy will contribute to the formation of rational and reasonable pattern of man-earth areal system in China[13,19]. Ecological protection and compensation will be incorporated into governmental functions of macro-regulation and coordination. Environment and ecological problems in surrounding area will have increasing impact on metropolitan development, which needs to be supported by ecosystem service in broader areas. Therefore, large cities and the surrounding area will become an integrated whole interweaved by eco-system service and ecological compensation. Ecological compensation and financial transfer payment will be a research topic for regional development scholars[45,46].

2.4 Scientific and Technological Demands in Regional Development (2030–2050)

From 2030 to 2050, economic growth will tend to decline and stabilize. According to the inverted U-shaped relationship between economic growth and regional imbalance, China's socio-economic spatial structure will become relatively balanced. However, the pressure of enormous economic aggregate and high level of urbanization on resources and environment at current stage will remain strong for a long time and the problems of spatial structure and organization will become more complicated. Most of the above important issues will continue to exist in China's regional development. Moreover, some new scientific and technological demands may emerge in the field of regional development.

2.4.1 To Ensure Homeland and Ecological Security at Various Scales

Ecological security pattern refers to the landscape elements that are crucial to maintain the health and security of ecological process, their location and inter-relations, such as mountains and water, wetland system, natural form of river system, greenway system, and the Protection Forest System in China established in the past *etc*. It is a multi-level, continuous, and integrated network composed of macro-level homeland and ecological security pattern, regional ecological security pattern, and micro-level, urban and rural ecological security pattern. Ecological security pattern at these different scales constitute ecological infrastructure safeguarding the security and health of our ecological system.

2.4.2 Cross-border Cooperation and Development

Under economic globalization, the cooperation between China and other surrounding areas will continue to be expanded and deepened, which is another important issue for national policy-making. In response to this trend and geo-political and geo-economic cooperation and conflicts, we need to constantly push forward regional cooperation in various fields, and to establish cross-border economic cooperation zones. Potential candidates include regions in northeast Asia, central Asia, the Lan Cang-Mekong river region, Sino-Indonesia cooperation region, East/Southeast Asia-Australia/Singapore cooperation region, and so on. We need to do prospective studies of these regions on such issues as current economic conditions, economic cooperation framework, the selection of economic center, and key basic infrastructure *etc*.

Energy will be a key field of cooperation between China and its neighboring countries in the future. The spatial pattern of China's energy security strategy can be preliminarily divided into north line, western line, and southern line. For north line, the establishment of land-based oil pipeline from Russia Angarsk to Daqing (Angarsk-Daqing Line) is of great significance in satisfying the oil and gas demands of Northeast China, North China, and East China. For western line, the resources of oil and gas in Central Asia are of great significance in satisfying the oil and gas demands of West China, especially Southwest China. In terms of middle to long term cooperation, Xinjiang Autonomous Region will be an important node of our country with imported crude oil and natural gas pipelines as well as cross-border railways. Southern line strategy refers to the oil cooperation between China and Southeast Asian countries, through which energy pipelines enter Southwest China such as Yunnan and Sichuan, *etc*.

Due to the increasing globalization of environmental problems and the barrier of national boundaries for environmental treatment, it is very important to establish a framework of global environmental governance to deal with global environmental problems. The construction of this framework will thus be listed on cross-border regional cooperation agenda.

Studies should be carried out on cross-border regional issues in order to serve national strategic interests. At the same time, international communication and cooperation should also be expanded. Despite the particularity of regional problems in China, they still have a lot in common with those in developed countries. Therefore, strengthening international communication and cooperation with other countries can broaden our horizon in regional research. In addition, as China's foreign investment are growing rapidly, it is urgent to conduct research on China's outward investments in order to obtain valuable information and get access to overseas strategic resources.

2.4.3 Low-carbon Economy Will Become New Growth Engine of Regional Economy in the Future

In order to prevent irreversible deterioration of global climate, we should promote the reduction of greenhouse gas emission, which provides both opportunities and challenges for the adjustment and upgrading of regional industrial structures. Low-carbon economy, based on low energy consumption and low pollution, is considered to be the fifth revolutionary wave following the industrial revolution and the information revolution that will dramatically reshape the world economy. New energy, clean energy, bio-energy, renewable energy, *etc.* are all part of the low-carbon economy. Against the background of global greenhouse gas emission reduction, the amount of carbon emission will be an important resource of regional development and regional economy based on low-carbon industries will have strong potential to achieve sustainable growth in the future. Furthermore, global greenhouse gas emission reduction will reshape the spatial pattern of China's international trade.

2.4.4 Regional Problems Caused by the Reconstruction of Global Financial System

International monetary system may go through changes. The current international monetary system is a kind of floating system centered on US dollar and diversified currency reserves. Its risk lies in that once US trade deficit cannot be sustained, US dollar will depreciate to a large extent, which in turn may bring about turmoil and crisis in global financial market. For trade surplus countries, rapid appreciation of exchange rate may weaken their export competitiveness.

In the long run, if the United States continues to rely on the savings provided by other countries and its share in global economy continues to decline, the status of US dollar as world's reserve currency will definitely be weakened, and the existing global monetary system will be fundamentally restructured. More fluctuations will occur in global financial market before the establishment of a new system, which will necessarily have a negative impact on the operation of global industrial economy. Such chaos has ever happened in the 1970s after the collapse of the Bretton Woods system. As exchange rate is the core component of financial system, any possible fluctuations of RMB exchange rate may significantly affect the future growth pattern of China's regional economy.

3 Major Research Issues and Roadmap Design in Regional Development Research

Based on the analysis of regional development trends and national scientific and technological demands in Chapter 2, we can understand that along with the development of economic globalization and global economic integration, the rapid process of industrialization and urbanization as well as the employment of powerful new technologies are profoundly reshaping the basic structures of man-earth areal system, the evolutionary process of regional development, and the spatial pattern of global and national development. Therefore, it is of urgent strategic importance and scientific significance to strengthen the studies of regional development in order to provide scientific support for central and local strategic policy-makings.

3.1 Scientific and Technological Development Objectives in Regional Development

3.1.1 The Identification of Major Research Fields

The main research fields of regional development include:

(1) Man-earth relationship—the evolving mechanism and analytical method of man-earth areal system, including the basic theories of man-earth areal system, the analysis and simulation of the development status of man-earth areal system, and the method for optimizing and regulating regional man-earth relationships at different spatial scales.

(2) Regional development—evolutionary theories of regional development patterns and regional governance, including regional responses to economic globalization and global environmental changes, research on the formation, transformation and differentiation of industrial space, and the trend of interregional interaction and its influence on regional development patterns.

(3) Urban-rural structure—urbanization process, the evolution of urban-

rural structure and its resource and environmental effects, including the driving mechanisms and spatial structure optimization of urban evolution, the resource and environmental effects of urbanization process and urbanization mode, the processes and mechanisms of spatial agglomeration and diffusion in metropolitan areas, and the evolution of urban-rural relationship and urban-rural integrated development mode.

3.1.2 Short-term (to 2020) Development Objectives

In view of the regional presence of contradictions among population, development, resources and environment during the 30 years of reform and opening-up, we should analyze and evaluate the key problems and trends of regional development in China, and make efforts to reveal the strategies and spatial processes of regional responses to globalization and urbanization. Meanwhile, we should also improve the spatial recognition system of territorial functions, develop both theories and methods for understanding the processes of inter-regional material and energy flow and the agglomeration and dispersion of population and industry, and establish data collection network and information platform for the support of dynamic monitoring of regional development status. Progress should be made in the following three fields:

(1) Study on the evolving mechanisms and simulative analysis of man-earth areal system: to reveal the dynamic mechanisms and evolutionary theories underlying the optimization of man-earth areal system, to develop analytical model of regional sustainable development status, and to form preliminarily simulation platform for the optimization and scenario analysis of regional man-earth relationships.

(2) Study on the evolutionary theories and regulation of regional development: to explore the mechanisms of regional development and the theories of spatial transformation, to put forward functional zoning schemes and sustainable development models for typical regions in our country, to develop both theoretical and methodological system for comprehensive evaluation of regional planning and other key projects, and to conduct in-depth research on the spatiality of new industries and the resource and environment effects of basic infrastructure, *etc*.

(3) Study on urbanization and the evolutionary mechanism of urban-rural structure: to illustrate the evolving trend of urbanization process and urban-rural structure, to reveal the driving mechanisms of urbanization and the resource-environment effects of spatial expansion, to study the relationship between the transfer of rural surplus labor force and land use change, and to put forward policies and measures to achieve coordinated development of rural and urban areas.

A review of the demands for regional development research

From international perspective, regional development research has become an important component of resource and environment science and earth system science against the increasingly overwhelming trend of economic globalization and the prominent problems of global environment. Strengthening the basic research on regional development and providing scientific support for central and regional policy-makings are favorable to the improvement of sustainable development capability and global competitiveness of regions in China and are vital to the future development of our nation and the achievement of the objective of constructing a harmonious society.

From national perspective, China is currently at a historical juncture characterized by the formation of socialist market economic system and significant transformation of societal institution. Under rapid urbanization and industrialization, regional development research needs: 1) to seek new growth trajectory of urbanization and industrialization with Chinese characteristics that fits into the national conditions of our country; 2) to build up a powerful science and technology support system, on the basis of regional development theories for the implementation of scientific outlook on development, "Five Integration" strategy, the construction of harmonious and well-off society, and the development of recycling economy and environment-friendly economy, *etc.*; 3) to provide strategic decision-making and action support for key regional development problems including solving the "three agricultural problems" confronting China's agriculture, rural areas and farmers, promoting the coordinated development between urban and rural areas, narrowing regional gap, accelerating the construction of industrial agglomerations, preventing environmental pollution, avoiding ecological degradation, optimizing industrial structure, speeding up industrial transformation, putting forward regional innovation, and promoting the intensive and effective utilization of land, *etc*.

On the regional level, the regionalization tendency of urban areas and the urbanization tendency of regions are two sides of the same coin in urban and rural development. Mutual interaction between them is increasingly blurring the urban-rural boundary and intensifying the conflicts and contradictions between urban and rural areas. Therefore, it is urgent for us to conduct regional and urban planning in different types and at different layers in order to solve conflicts, reach agreement, eliminate plundering behavior, and achieve common prosperity. Regional development research is urgently needed to cope with the sustainable development issues in different types of functional areas, such as natural reserves, ecologically fragile areas, areas with severe water shortage, natural disaster ridden areas, key development areas, areas undergoing optimization and integration, industrial agglomeration areas, infrastructure areas, and development-restricted areas, *etc*. which can further enhance regional competitiveness, independent innovation ability and sustainable development capability.

3.1.3 Medium-term (to 2030) Development Goals

To meet the national strategic needs, we need to perfect the theoretical and practical system of regional development in our country, to study the natural foundation, material security, economic and social process, and the eco-environmental effects of regional sustainable development, to construct functional areas according to the principle of coordinated regional development and figure out the regional planning and regional policy support system of these areas, to develop different sustainable development models for different types of regions and bring the guiding role of these models into full play in practice. In addition, we should take the study on the spatial pattern and evolutionary process of humanistic elements as the main stream to achieve scientific understanding of the mechanisms of new factors in regional development and illustrate the spatial-temporal process, mechanism and the social, economic and environmental effects of related policies underlying the evolution of man-earth relationship. Furthermore, we should construct man-earth areal system development theory and the theoretical system and integrated analytical methods of, regional development and urbanization, illustrate the theory of industrial spatial differentiation and the theories of urban and rural development and inter-regional interaction, which all aim to establish the leading role of human factors in earth system science research. Last but not least, we should study the impact of key strategic industries, new industries and regional integrated infrastructure on regional spatial differentiation, reveal the mechanism of regional sustainable development and its evolutionary theory and regulation approach, explore the regional differentiation theory of agricultural development in China and the approach to achieve coordinated development of urban and rural areas, and establish the simulative analysis platform of the evolutionary status of man-earth areal system.

3.1.4 Long-term (to 2050) Development Objectives

Our long term development goals are to focus on the study of the driving factors, the evolutionary mechanism, the evolution of spatial pattern, development models, changing economic and technological parameters, and the integration theory of regional development through the analysis of process and mechanism, simulative evaluation, and experimental and demonstrative research, to explore the development theory, structure, and governance regime of the "territorial system of human activities" in the earth surface science system, to investigate the role of human factors and their resource and environmental effects, as well as the evolutionary theory of "man-earth areal system" under human interference. We should stick to meeting national strategic demands as the starting and ending point, continuously trace and carry out prospective study on the driving factors, resource-environment foundation, development theories and scientific models of regional development and urbanization process in China, explore the scientific mechanism of and technical approaches to the achievement of the social and economic spatial structure characterized by the rational combination of economization, harmony and efficiency, and provide mature scientific and strategic support for our country through macro-level strategic design, meso-level research on pattern, driving force and mechanism, and micro-level experimental and demonstrative research.

3.2 Roadmap Design for Scientific and Technological Demands in Regional Development Research

3.2.1 Science and Technology Framework

This book will design the Roadmap of China's scientific and technological demands in regional development research based on the scientific understanding of regional development theories and the characteristics of the natural ecosystems and socio-economic system of our country, with an aim to meet major national strategic demands and forecast the research frontiers in this field. To this end, the basic modules of regional development research and their inter-relationships are shown in Fig. 3.1.

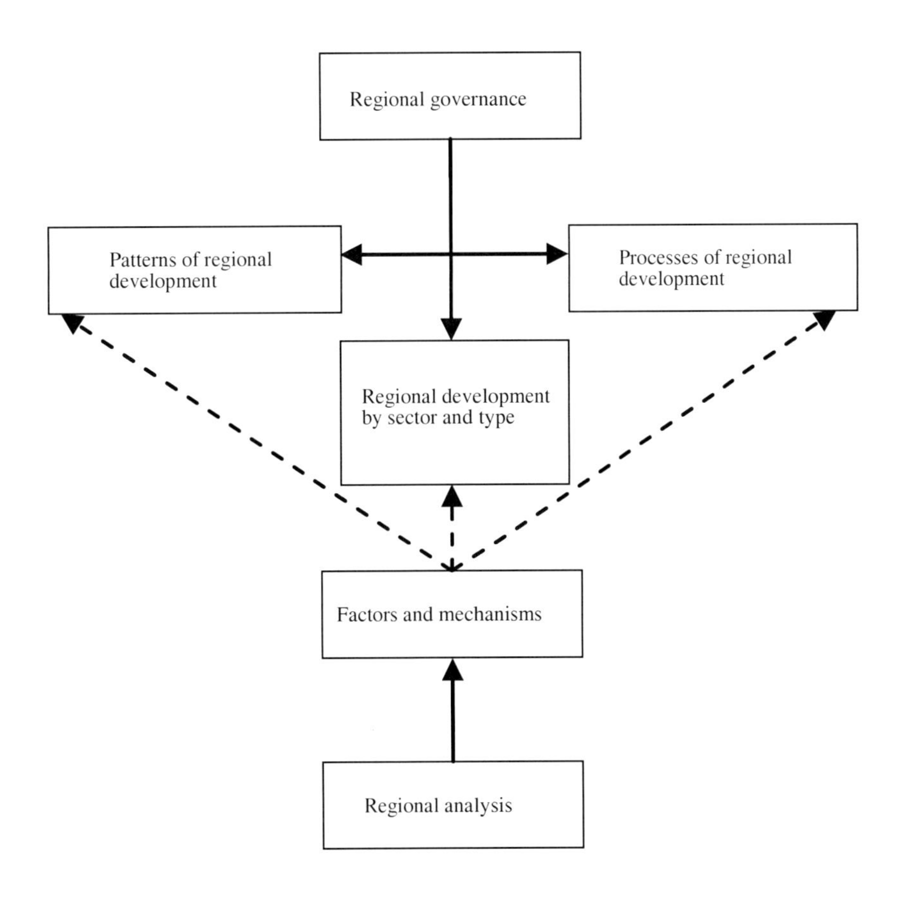

Fig. 3.1 The basic modules of regional development research

With the support of modern regional analytical methods and techniques, we should probe into the affecting factors of regional development and their operational mechanisms, conduct integrated research on the spatial distribution theories of various socio-economic sectors and the spatial organization theories of different types of regions, generalize the formation and spatial-temporal transformation theories of regional development patterns, and propose policies and measures for the regulation of human spatial behavior and territorial function.

3.2.2 The Roadmap of Scientific and Technological Development

According to the overall objective and the basic module design of regional development, the patterns and processes of regional development constitute the long-term focus of the scientific and technological demands of regional development research. More specifically, it will reveal the formation and spatial-temporal differentiations of regional development patterns based on new factors, mechanisms, sectors, and regional types in different time in order to achieve the sustainable development in the science and technology field of regional development. Recent research in such field should focus on the factors and mechanisms of regional development, the theories of spatial distribution of different sectors and different types of regions which can provide basic theoretical support for regional development research and facilitate the solving of major practical problems in China's regional development. The medium-term goals are to make a breakthrough in the research methodology of regional governance and regional research, improve the basic theoretical system of regional planning, enhance the technological level of regional development research in our country, and improve the capability of regional development research to serve practical and national scientific and technological demands.

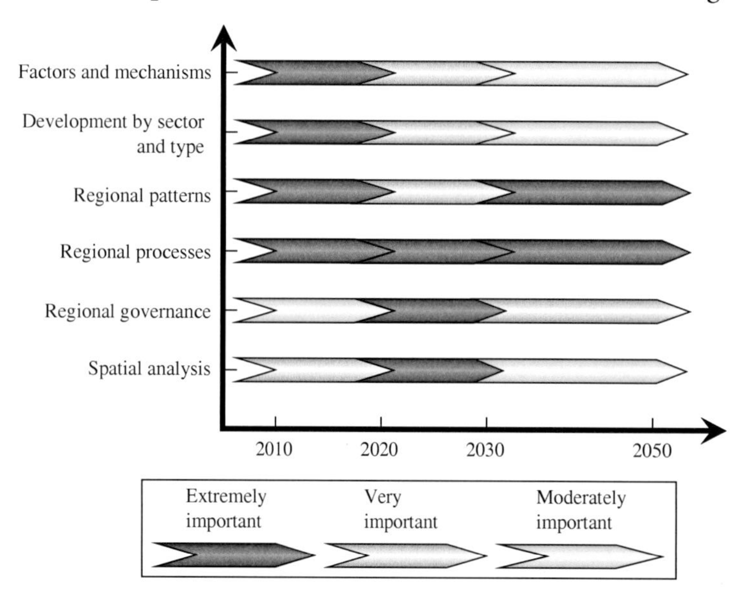

Fig.3.2 Roadmap design of regional development research at different stages

3.2.3 The Roadmap Design of Scientific and Technological Objectives in Regional Development Research

Main research directions will be presented based on a further decomposition of the above 6 basic modules. The scientific and technological objectives of regional development research will be determined according to the scientific and technological demands of our country in several major directions, which include:

(A) The Research Field of "Factors and Mechanisms"

Main research directions include the study of old and new factors and mechanisms.

(1) In the research on old and new factors, major national demands involve the identification of physical and human elements which can influence the prospects of regional development in China, the exploration of major affecting factors at different development stages and in different regions, the study of the trends of China's regional development under the influence of both old and new factors, and the development of policies and measures for responding to the changes of factors and promoting regional development. Therefore, the main scientific and technological objectives of regional development research are to study the interactive relationships among traditional factors including natural resources, ecological environment, and infrastructure systems, *etc*., and the impact of such interactions on the spatial patterns of regional development, to investigate the degree, manner and effects of the impact that new factors such as climate change, informatization, globalization, technological innovation, social culture, *etc*. have on the spatial organization of regional development, and to probe into the role of human being in man-earth areal system.

(2) In the research on mechanisms, major national demands involve the investigation of basic theories guiding the direction and process of China's regional development, the formulation of coping strategies in response to the differences in the development mechanisms of regions characterized by different conditions, levels and structures, and the design of evolving trajectories to achieve coordinated, healthy, and sustainable regional development. Accordingly, the main scientific and technological objectives of regional development are to study the driving forces and mechanisms of regional development in China and the scientific basis of the differentiation of regional development models.

(B) The Research Field of "Development by Sector and Type"

Main research directions in this area include the study of the development of various industrial sectors, population and social culture, and different types of regions.

(1) In the research on development by industrial sectors, major national demands include the investigation of the comparative advantages and evolutionary processes of various industrial sectors, the changes in the spatial

distribution of different industrial sectors, and the path and preconditions to achieve regional industrial upgrade. Accordingly, the main scientific and technological objectives of regional development research are the formation and transformation process of industrial space, the evolution process and benefits of territorial division of labor at different spatial scales, and the spatial distribution theory of new types of regional economy.

(2) In the research on the development of population and social culture, major national demands include the formulation of strategies to cope with the emerging trend of population growth and flow, to optimize the spatial distribution of social and cultural facilities, and to draw up both objectives and measures for the construction of soft environment favorable to regional development. The main scientific and technological objectives in this regard are the research on the impact that the driving force of regional development potentials have on the spatial distribution of population and the operational mechanism underlying the impact of soft environment and basic public services on the changes of regional development potentials.

(3) In the research on different types of regions, major national demands involve the scientific judgments and prospective forecast of the development conditions of regions featured by different physical types and development levels, especially those problematic areas and vulnerable places,the formulation of regional policies to promote the development of different types of regions, and the research on the possible effects of non-physical or non-conventional space utilization on the patterns of regional development. Therefore, the main scientific and technological objectives of regional development in this regard include the index system, measurement methods and regulation technologies of regional development conditions, status and prospects, and the regional effects of spatial structure evolution in the process of spatial expansion.

(C) The Research Field of "Regional Patterns"

In this subfield, the researches are mainly concerned with regional development on a global scale, the status and patterns of China's regional development, urban-rural interaction, and the inter-regional relations.

(1) In the research on regional development at a global scale, major national demands involve the impact of globalization on China's regional development and related coping strategies, the preconditions for speeding up the regional integration process between China and its neighboring countries, and the approach to strengthen China's global influence and control power. Accordingly, the main scientific and technological objectives of regional development are to understand the mechanism of regional response to globalization, and the characteristics of cross-border inter-regional interaction.

(2) In the research on the status and patterns of China's regional development, major national demands include the evolving trends of man-earth relationship in different regions of China and our countermeasures, future patterns of regional competitiveness in China, measuring the evolving trend of the regional disparity in China, and the policy framework for reversing such

trend. Therefore, the main scientific and technological objectives of regional development are to examine the natural foundation, material security, socio-economic processes and eco-environmental effects of regional sustainable development, the structural characteristics and evolutionary process of man-earth areal system, the "environmenta-social dynamics" of man-earth relationship, and the general theory of spatial structure evolution of regional development in China.

(3) In the research on urban-rural interaction, major national demands involve the solutions to eliminate urban-rural opposition and promote urban-rural integration, and the measures for solving land use conflicts and coordinating urban and rural functional orientation in the process of urban expansion. Thus, the main scientific and technological objectives of regional development are to investigate the internal mechanism, external conditions and index system of urban-rural integration.

(4) In the research on inter-regional relations, major national demands include the trends and cost-benefit of large-scale inter-regional flow of water and mineral resources, energy, *etc*., the compensation mechanisms and other institutional arrangements in the process of economic interests transfer among different functional regions, design of policy system in terms of ecological compensation, clean energy quota, and environmental capacity (pollutant emission right), *etc*. Accordingly, the main scientific and technological objectives of regional development are to develop the theories of the formation and transformation of inter-regional flow of material, energy, population, information, technology, finance, *etc*., and the theory of inter-regional interaction under the influence of new factors and new mechanisms, to calculate the costs and benefits of inter-regional interaction in open system, and to evaluate the rationality and feasibility of the main approaches to strengthen basic-level, inter-regional connections at micro scale.

(D) The Research Field of "Regional Process"

In this subfield, main research areas include regional strategies in response to global climate change, the processes of globalization and regional integration, the processes of industrialization and urbanization, and the processes of functional differentiation and regional equilibrium.

(1) In the research on regional strategies in response to such key process as global climate change, major national demands involve the basic strategies and regional modes for the formation of low-carbon economy and social system, the costs and benefits of China in terms of carbon emission in the process of industrial transfer and international trade, and the market structure of inter-regional carbon emission trading and its regional effects. Accordingly, the main scientific and technological objectives of regional development are to analyze the objectives, index system, and regulation mechanism of regional economic growth in three-dimensional objective space.

(2) In the research on globalization and regional integration process, major national demands include the analysis of both positive and negative

effects of globalization on China's regional development and related policy design, of the strategies and focus areas of regional cooperation between China and its neighboring countries, and of the position of regional integration centered on China's eastern coast in globalization process. Therefore, the main scientific and technological objectives of regional development are to provide scientific judgments and theoretical explanation of the effects of economic globalization, and to analyze the formation conditions and spatial structure of regional economic integration and its impact on China's regional development.

(3) In the research on industrialization and urbanization processes, major national demands include the judgments of their future development, regional models, and the resource and environmental effects. Thus, the main scientific and technological objectives of regional development are to investigate the driving forces of industrialization and urbanization in China, to conduct comprehensive cost-benefit evaluation of the two processes, and to reveal their regional differentiations.

(4) In the research on functional differentiation and regional equilibrium, major national demands involve the basic trends of population flow and the spatial distribution of industry, the formation conditions and future construction arrangements of regions for ensuring ecological and food security, the spatial organization of basic public service system, and the evolving trajectory of regional development gaps. Accordingly, the main scientific and technological objectives of regional development are to investigate the forces underlying the formation of territorial functions and their evolution, the basic theory and model representation of regional equalization process, and the threshold conditions and cost-effectiveness of the widening/narrowing of regional development gaps.

(E) The Research Field of "Regional Governance"

In this field, main research areas are spatial planning and regional policies.

(1) In the research on spatial planning, major national demands involve the establishment of spatial planning system, the integration of spatial planning resources, the formation of theoretical and methodological system of spatial planning, the cultivation of professional and technical talents, the establishment of spatial planning institutes, and the creation of favorable institutional environment for the formulation and implementation of spatial planning.

(2) In the research on regional policies, major national demands include establishing regional policy system based on classified management, and improving institutional arrangements for the effective implementation and evaluation of regional policies. Therefore, the main scientific and technological objectives of regional development are to optimize the allocation of resources of regional development that are controlled by the government, to analyze the mechanism for the regulation of vertical or horizontal spatial organization of regional development (including spatial planning and regional policy system), and to establish the scientific system of spatial planning and regional policies.

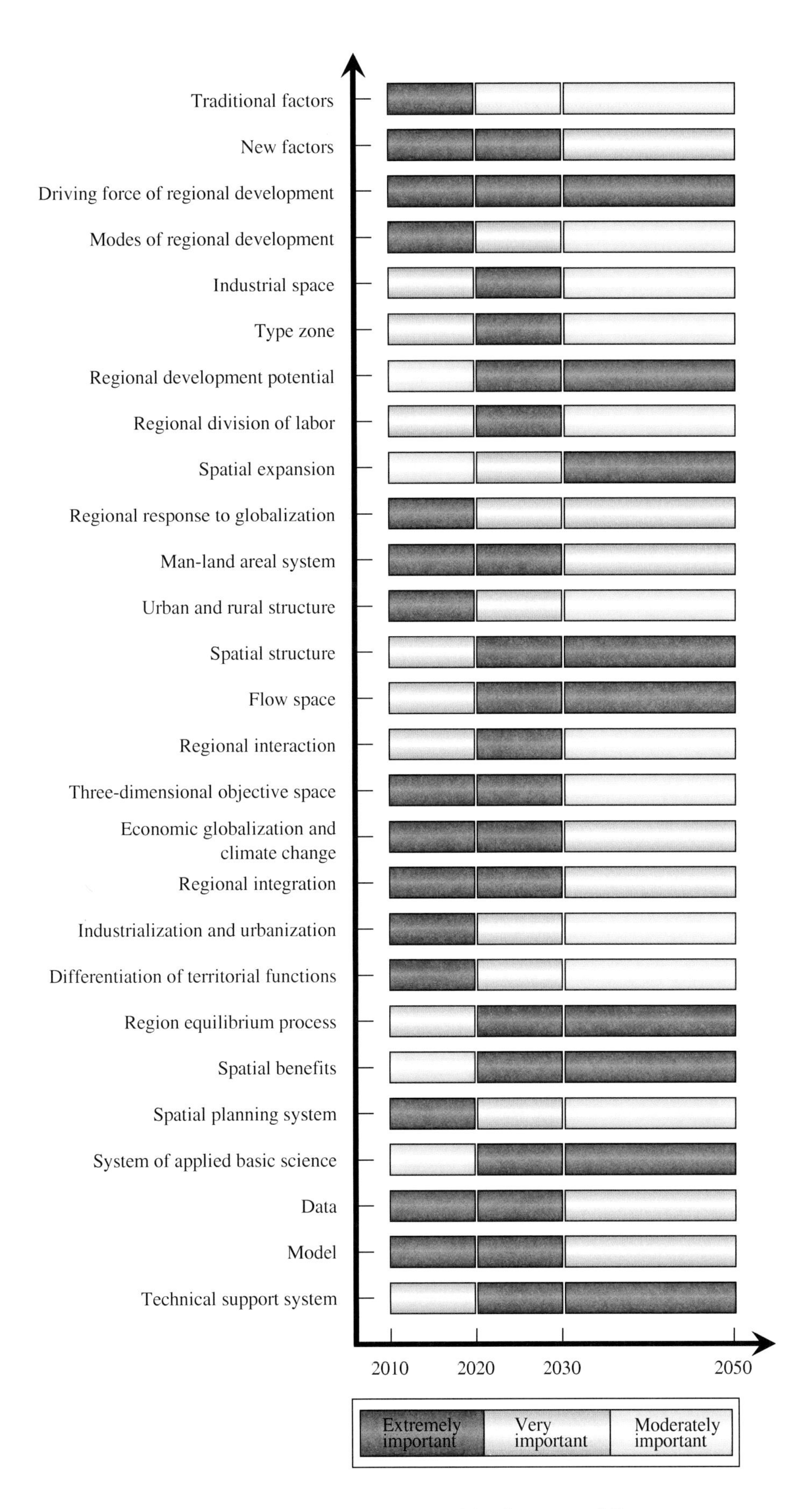

Fig.3.3 Major research questions of regional development at different stages

(F) The Research Field of "Regional Analysis" (Research Methods of Regional Development)

In this subfield, main research areas include data acquisition and management, spatial analytical techniques, and prediction, simulation and optimization.

(1) In the research on data acquisition and management, major national demands and scientific and technological objectives involve the access to first-hand information, new technologies (*e.g.* remote sensing) and statistical data, database construction in support of regional development research and decision-making, and the improvement of acquisition capability of real-time, comprehensive, and dynamic data.

(2) In the research on spatial analytical techniques, major national demands and scientific and technological objectives include developing spatial analytical models of regional development, characterizing the basic conditions of regional development, and achieving visual expression of spatial analytical process.

(3) In the research on prediction, simulation and optimization, major national demands and scientific and technological objectives are to model regional development process and prospect, and optimize the technical support system of regional development.

3.2.4 Strategic Arrangements of Key Research Questions at Different Stages

Table 3.1 The key research questions of regional development and their phased-in arrangements

(Continued)

Scientific and technological fields	Research scopes	Key research questions and development strategies by stage			
		Key research questions	Short-term	Medium-term	Long-term
The factors and mechanisms of regional development	Traditional factors	1) The role of physical-geographical and environmental factors in the evolution of industrial structure and its spatial-temporal differentiations	√		
		2) Interactive relationships between regional development and the affecting factors such as resources, environment and disasters	√	√	
		3) Operational mechanism of geographical location and basic infrastructure system on the changes of regional development patterns	√		
		4) The role of traditional factors in determining the location advantages and its impact on the patterns of regional development around the world	√	√	

(Continued)

Scientific and technological fields	Research scopes	Key research questions and development strategies by stage			
		Key research questions	Short-term	Medium-term	Long-term
The factors and mechanisms of regional development	New factors	5) The mechanism of regional response to economic globalization and climate change	√	√	
		6) The mechanism underlying the impact of informatization and motorization processes on regional development patterns	√	√	√
		7) The manner and extent of the impact that technological innovation has on regional development	√	√	√
		8) The mechanism of the impact of social and cultural factors on regional development		√	
		9) The mechanisms of the impact of new factors on the spatial distribution of different industrial sectors and the development of different types of regions	√	√	√
	Driving force of regional development	10) Cost-benefit analysis of economic interests in regional development process	√	√	
		11) Influence of human development values and consumption orientation on regional development objectives and paths		√	
		12) Influence of ethics and values on regional development modes			√
		13) Basic theory of the evolution of China's regional development	√	√	
	Modes of regional development	14) Interactive mechanism, territorial differentiation and the spatial-temporal evolution of regional resource-ecology-economy system	√		
		15) The theoretical basis of regional development modes in China	√	√	
Sector and type zone development	Industrial space	16) Life cycle theory of industrial space	√		
		17) New patterns such as advantageous production zones and industrial belts, *etc.* in agriculture and the regional differentiation theory of agriculture	√		

(Continued)

Scientific and technological fields	Research scopes	Key research questions and development strategies by stage			
		Key research questions	Short-term	Medium-term	Long-term
Sector and type zone development	Industrial space	18) Causes and impacts of the new trend of industrial distribution	√		
		19) The spatial structure of central function and service system of modern service industry and its impact on regional patterns	√	√	
		20) Sustainable utilization of tourism resources and the spatial organization of leisure industry	√		
		21) Mutual interaction of industrial spaces and the formation and evolutionary process of industrial space system		√	
	Type zone	22) The index system and calculation method of the development status of different type of regions	√	√	
		23) Ways to enhance the influence and control power of developed areas	√		
		24) Ways to enhance the sustainable development capacity of underdeveloped regions	√		
		25) Causes underlying the formation of problem areas and the principles of reconstruction	√	√	
		26) Spatial distribution principles of new types of regions	√	√	√
		27) Regional policies based on classified management and their scientific basis	√	√	
	Regional developme potential	28) Driving mechanism of population growth and flow	√	√	
		29) The construction process and operational mechanism of soft environment favorable to regional development		√	√
		30) Regional competition theory and the ways to cultivate regional competitiveness	√	√	
		31) The methods and regional effects of industrial optimization and upgrade	√		

(Continued)

Scientific and technological fields	Research scopes	Key research questions and development strategies by stage			
		Key research questions	Short-term	Medium-term	Long-term
Sector and type zone development	Regional division of labor	32) The changes of location factors of main industrial sectors and their impact on regional comparative advantages	√		
		33) Evolutionary process and cost-effectiveness of spatial division of labor at different spatial scales	√	√	
		34) The conditions and scenarios of the restructuring of international division of labor		√	√
	Spatial expansion	35) The impact of virtual space created by the Internet on the spatial distribution of physical space	√	√	
		36) The enhancement of the competitiveness and sustainable development capability of non-physical spaces such as society and culture		√	√
		37) Prospects of the utilization of non-conventional spaces such as the outer space, underground and the ocean, *etc*. and their impact on the changes of regional development patterns			√
		38) Spatial structure changes in the process of spatial expansion and their regional effects	√	√	√
Regional patterns	Regional response to globalization	39) Impacts of globalization on China's regional development and related coping strategies	√		
		40) Conditions and paths for speeding up the process of regional integration between China and its neighboring countries	√		
	Man-earth areal system	41) Judgment standard and the objectives of the status of man-earth relationship and regional sustainable development		√	
		42) The improvement of the supporting capacity of resource and environmental security guarantee system in regional development	√		

(Continued)

Scientific and technological fields	Research scopes	Key research questions and development strategies by stage			
		Key research questions	Short-term	Medium-term	Long-term
Regional patterns	Man-earth areal system	43) Location selection of socio-economic system and the spatial coupling processes among the regional patterns of land use and ecological function		√	
		44) Paths of the optimization, regulation and integration of man-earth relationship at different spatial scales	√	√	√
		45) Structures and functions of man-earth areal system	√	√	
	Urban and rural structure	46) Basic characteristics and trends of China's urban-rural relations	√		
		47) The theories of urban-rural interactions, their effects and the trend and countermeasures for coordinated development of urban and rural areas	√	√	
		48) Regional patterns and socio-economic mechanisms of new rural construction	√		
		49) The regional pattern of urban agriculture and the mechanism of urban-rural coordination	√		
		50) Structural evolution and optimization approaches of city, town and village system	√	√	√
	Spatial structure	51) Evolutionary theory of the "Point -axis-area" spatial structure and its application	√	√	√
		52) Optimizing the combination of living space, production space and ecological space on the earth surface	√	√	√
	Flow space	53) Interactive modes of the material flow of man-earth system		√	√
		54) Formation mechanism of the space of flow and its impact on regional development patterns		√	√
		55) New features of population flow and the optimization of its regulation	√	√	

(Continued)

Scientific and technological fields	Research scopes	Key research questions and development strategies by stage			
		Key research questions	Short-term	Medium-term	Long-term
Regional patterns	Flow space	56) The formation mechanism, support system distribution and environmental effects of large-scale inter-regional material flow	√		
		57) Node distribution of the space of flows such as information, technology, and finance, *etc*.	√	√	
	Regional interaction	58) The change of the structure, intensity and direction of inter-regional flows and their carriers	√	√	√
		59) Efficient and intensive inter-regional interactions of material, energy , information and technology and their synergistic effects on regional development		√	√
		60) Causes of the changes of inter-regional interaction theory and the new expressions	√	√	√
Regional process	Three-dimensional objective space (economic growth, social welfare and ecological environment)	61) Mechanism and trend of regional development in the three-dimensional objective space	√	√	√
		62) Regional smart growth theory and simulative analysis methods	√	√	
		63) Forecast technologies of regional development status	√		
		64) Index system of the early-warning systems of Regional development	√	√	
	Economic globalization and climate change	65) The trend and positive and negative effects of economic globalization and its effects on global development	√		
		66) Evolutionary theory of the regional pattern of China's open-up and international trade	√	√	
		67) Opportunities and countermeasures in the process of global industrial restructuring	√	√	√
		68) Potential assessment and cost-benefit accounting of low-carbon industrial structure in supporting sustainable regional growth		√	

(Continued)

Scientific and technological fields	Research scopes	Key research questions and development strategies by stage			
		Key research questions	Short-term	Medium-term	Long-term
Regional process	Economic globalization and climate change	69) Comprehensive balancing methods for the costs and benefits in the spatial restructuring of China's foreign trade and industrial transfer under the influence of global greenhouse gas emission reduction	√		
	Regional integration	70) Basic pattern and contrasting relations of global regional integration		√	
		71) Geo-political and geo-economic explanations of the formation of cross-border regional economic cooperation bloc	√		
		72) Strategic emphasis and steps of regional cooperation between China and its neighboring countries	√	√	
	Industrialization and urbanization	73) Driving mechanism of urban evolution and industrial structure upgrade	√		
		74) Scientific basis of the objective, speed and scale of urbanization	√		
		75) Spatial organization and mechanism of new industrialization		√	
		76) Optimization methods of the spatial patterns of urbanization and industrialization	√	√	
		77) Process and mechanism of spatial agglomeration and diffusion of urban areas	√		
		78) Formation of industry agglomerated areas and the improvement of their competitiveness	√		
		79) Regional model selection and the resource and environmental effects of industrialization and urbanization	√	√	
		80) The index system and management of the resource-saving and environment-friendly model of urbanization and industrialization	√	√	

(Continued)

Scientific and technological fields	Research scopes	Key research questions and development strategies by stage			
		Key research questions	Short-term	Medium-term	Long-term
Regional process	Different-iation of territorial functions process Differentiation of territorial functions	81) Theory of the formation and evolution of territorial functions	√	√	√
		82) Identification methods and zoning techniques of territorial functions	√		
		83) Balance mechanisms and institutional arrangements of the economic interests transfer among different functional regions		√	
		84) Technologies for the dynamic monitoring and comprehensive evaluation of the evolution of territorial functions	√	√	
	Regional equilibrium process	85) Basic trend and mechanism of the spatial process of agglomeration and diffusion of population and industry	√	√	
		86) Basic conditions and regulation objectives of the formation and transformation of regional disparity (differences)	√	√	
		87) Standards and spatial allocation rules of basic public service system		√	√
	Spatial benefits	88) Calculation methods and theoretical basis of both long-term and comprehensive benefits of spatial structure		√	√
		89) Main factors of highly efficient spatial organization and the real method of implementation		√	√
Regional governance	Spatial planning system	90) Approaches and technical methods of spatial structure optimization	√	√	
		91) Objectives, contents and mutual coupling of major types of spatial planning	√		
		92) Evaluation methods of the effects of the implementation of spatial planning		√	

(Continued)

Scientific and technological fields	Research scopes	Key research questions and development strategies by stage			
		Key research questions	Short-term	Medium-term	Long-term
Regional governance	Scientific system of the application basis of spatial planning	93) The geographical foundation of spatial planning	√		
		94) Basic theoretical system of the spatial organization of regional development		√	
	System of applied basic science of spatial planning	95) Technical methods of spatial planning such as computer and mapping technologies *etc.*	√	√	
Spatial analysis	Data	96) Multi-channel data acquisition, integration and database construction	√	√	
		97) Method of real-time and dynamic data acquisition and processing	√	√	
		98) Data-sharing mechanisms and efficient service system among different countries and sectors		√	
	Model	99) Application of the theories and methods of the open, complex and dynamic giant system in the research on the scientific and technological issues in regional development		√	√
		100) Technical methods for the transformation from micro-level conclusions to general theories		√	√
		101) Scenario analysis of regional development	√	√	
		102) Support system for the optimization of decision-making in regional development	√	√	
	Technical Support System	103) Development of informatization and intelligence in the research on man-earth areal system			√
		104) Dynamic simulation of regional development process based on visual expression		√	√

3.3 Transition of the Scientific and Technological Field in Regional Development Research

3.3.1 Regional Development in the Three-dimensional Objective Space

Regional development in the three-dimensional objective space refers to the development under the interaction among economic growth, social security and eco-environment. Social development at that time involves not only social security, but also the satisfaction of comprehensive human needs in all-round way. During such developmental process, the societal pressure on eco-environment will first increase gradually but then decline little by little. In this sense, both theories and analytical or simulative methods of regional economic development need to be reconsidered, while entails the coordination of three dimensions of economic growth, social development and eco-environment construction in regional development.

3.3.2 Regional Development under Three Requirements

The first requirement is to emphasize the livelihood issues in regional development, which focuses on the equalization of public services and people's living standard according to people-centered principle. The second requirement is to strengthen regional competitiveness through giving full play to regional comparative advantages, transforming economic growth models, and upgrading industrial structure on the condition that economic laws and market rules are followed. The third one is to achieve sustainable development of both domestic livelihood and international competitiveness based on the respect for natural law.

3.3.3 Regional Development in Three Spaces

The so-called three spaces refer to physical, virtual and unconventional spaces. The advent and development of the virtual space based on information network not only provides important technical carrier for the research on physical space, but also puts forward many new propositions regarding the virtual space in regional development, which in turn profoundly reshapes the patterns of physical space. Many new spatial forms and new factors and mechanisms that have impact upon the patterns of regional development such as research on cross-border regions will appear in the physical space. Moreover, the degree and prospects of the exploitation and utilization of unconventional spaces including the non-physical space composed of cultural and social phenomena, underground space, ocean space, and even the future aerospace will exert unpredictable influence on the regional patterns of physical space.

3.3.4 Regional Development Research Based on Three Integrations

Firstly, the existing integration of economic geography and resource and environmental sciences should be further strengthened. Secondly, emphasis should be placed on the integration with Data-Calculation-Simulation-Analysis technologies. It will be difficult to achieve the analysis, simulation, retrospective evaluation and early warning of the interactive relationships among urbanization, energy, water and soil, and foods, *etc*. without necessary technical means. It will also be impossible to bring the role of technology into full play in the policy-making and management process of regional development without practical data, timely reflection, visible expression, and the integration methods of micro and macro views. Thirdly, the integration with economics-based social sciences should be deepened and the interdisciplinary advantages should be brought into full play. The integration of society-economy-ecology system cannot be achieved without the contribution of Economics.

4 Integrated Propositions of Regional Development Research

Based on the scientific propositions of regional development research and in order to facilitate the solvement of key research questions at different stages, we choose the following three key integrated propositions that will last for five years as the guidance on the arrangement of key research projects in the field of regional development before 2020 under the support of major scientific programs in China such as the National Key Technology R&D Programs, Key Projects of National Natural Science Foundation, *etc.*

4.1 Theory of Territorial Function Formation and Regional Development Patterns in China

4.1.1 Significance of the Topic

The spatial differentiation of the territorial system on earth surface is the basic law of Geography and the basic perspective of geographical research. The territorial units with differentiated functions can form orderly organizations which constitute such spatial structures as von Thunen's circular structure and Christaller's hexagonal structure[1]. The difference of functional attributes is vital to territorial differentiation, while the scientific identification and rational arrangement of territorial functions are the prerequisite for the development of safe, effective and sustainable homeland space[13]. Traditional zoning is widely used in the research on territorial differentiation, such as the earlier forms of climate zoning, plant and animal zoning, comprehensive physical zoning, and the later forms of agricultural zoning and economic zoning, *etc.*[13,41-46]. However, there exists a series of difficulties and problems in the traditional zoning research. For instances, the comprehensive zoning with a combination of natural and human factors is still immature without clear explanations of the formation and evolutionary processes of territorial functions. These practical problems have significantly affected our further pursuit of zoning work, the process of discipline construction and even the achievement of the objective of meeting national strategic demands.

Foreign scholars have paid much attention to both theoretical and practical research on territorial function oriented zoning. For example, it is clearly stated in the *New Jersey State Development and Redevelopment Plan* that development zones, restricted areas, agricultural space, protected areas and other regions should be clearly delimitated in the planning. Territorial function oriented zoning has become an important tool for the rational development of land space and the formation of orderly spatial structure, which is supported by regionally-specific policies and measures such as "Spatial Encouragement", "Spatial Admission" and "Spatial Restriction", *etc.* However, compared with other countries, the research and practice of territorial function oriented zoning in China is relatively underdeveloped, which to some extent results in severe problems of regional disparity. More specifically, some regions push forward local industrialization and urbanization regardless of their resources and environmental carrying capacity and conditions. Consequently, various eco-environmental problems such as the dramatic decrease of green space including cultivated land, forest land and wetland, the severe over-utilization of underground water and surface subsidence, *etc.* emerged. It is urgently needed to reshape the pattern of territorial functions in China in response to the increase of regional resource-environmental pressure, the imperative to improve China's global competitiveness, and the gradual formation of new consumption concepts and scientific outlook on development[13,19].

Research on the formation and future evolution of territorial functions in China is not only an important practical field for theoretical and methodological innovation, but also the basis to meet national strategic demands in terms of the scientific design of territorial zoning function oriented schemes and the reorganization of national territorial space. Moreover, this program will exert significant and positive influence on the improvement of regional governance capability at various levels, the development of monitoring, evaluation and decision-making support system of functional zones, the coordination of territorial function oriented zoning and other departmental planning, and the strengthening of technical specifications and practical application of functional zoning methodology in different types of territory.

4.1.2 Research Objectives

(1) To explain the formation mechanism and future evolving trends of the territorial functions in our country, explore the basic methods of territorial function identification, and reveal the nature of different functional zones with a view to promote theoretical innovation on the territorial function differentiations on the earth surface system and to better meet national strategic demands of reshaping our homeland space.

(2) To build up the indicator system and hierarchical structure of territorial function oriented zoning, and explore the coupling relationships among different scales of territorial function zones.

(3) To make technological breakthrough in terms of the application of remote sensing, geographic information system in indicator selection, determination of zoning units, and dynamic monitoring and evaluation of zoning functions in territorial function oriented zoning, and to conduct research on the development of decision-making support system and information platform for territorial function oriented zoning, dynamic monitoring and supplementary analysis.

(4) To conduct preliminary research on the indicator system and technical specification for identifying territorial functions.

(5) To make substantial progress in data acquisition and management and coordinated development of China's functional zones with neighboring countries through project research.

In conclusion, this program will lay the foundation for the theoretical development of the orderly evolution of spatial structure and the spatial equilibrium of regional development, and provide theoretical basis and technical support for the improvement of our planning work. It is a comprehensive project of "data and zoning—regional develop patterns and processes—spatial structure and organization—trace, monitor, evaluation and spatial planning" that is supported by scientific methods, focused on regional development patterns at national-level and other related scales, and is characterized by solid theoretical foundation and broad scope of application.

4.1.3 Main Contents

(1) Investigation of the differentiations of territorial function and data platform construction. According to the requirements to identify territorial functions, comprehensive investigation of resources, ecology, environment, disaster, population, industry, urban and rural areas, society and infrastructure, *etc*. will be carried out in concerned regions in China and surrounding countries, as well as in important geographical units and along typical geographical lines. Integrated analysis and utilization of both geographical data and land and resource data that are closely related to the indicator system of territorial function oriented zoning will be conducted in order to achieve unified management, comprehensive analysis and visual expression of multi-scale, multi-type and multi-period data. In addition, visualized management system for the whole planning and implementation process of territorial function oriented zoning will be developed, and distributed and centralized information sharing platform of territorial functional zones will also be established.

(2) Basic research on the formation and evolution of territorial functions. The program will develop the spatial equilibrium model of regional development, characterize in an integrated way the evolving process of territorial functions through the calculation and illustration of territorial

function differentiations in major areas and based on regional research theories by sector and the distribution of geographic phenomena, and further discuss the evolving process and spatial-temporal differentiations of territorial functions through national and local case studies.

(3) Research on the methodology of territorial function identification and functional zoning. According to the spatial-temporal differentiations of the evolving process of territorial functions, this program will put forward the indicator system, individual index and integration algorithm, and assistant technical means for territorial function identification and territorial function oriented zoning, and reveal their coupling methods with the spatial structure of regional development. It will also conduct research on functional transformation at different spatial scales, analyze the measurement of important parameters and expected regulation approach, improve the tools of macro-control, and come up with policy suggestions for related planning.

(4) Research on the dynamic monitoring and retrospective evaluation of territorial function zones. This program will build up the monitoring and evaluation indicator system of territorial function zones through remote sensing, and make a breakthrough in the key technologies of remote sensing retrieval and dynamic monitoring and retrospective evaluation of the indictors of territorial functions. It will also comprehensively analyze and evaluate the direction and extent of changes in territorial functions and the underlying causes by making use of information-sharing platform, first-hand data and statistical data, and remote sensing data. Moreover, it will develop a software package of remote sensing retrieval of indicators and dynamic monitoring and evaluation based on multi-source data, and establish the operating system supporting the dynamic monitoring and retrospective evaluation of territorial function zones.

(5) Research on the decision-making support system and system integration of territorial function oriented zoning. This program will develop spatial analytical and simulative system, and the software package of dynamic monitoring, evaluation, management and decision-making support for territorial function zoning. It will also develop supporting technologies for integrating multi-sector and multi-level territorial function oriented zoning and different types of spatial planning. Moreover, it will develop decision-making support system for regional policy formulation and simulation analysis, scenario forecasting of the spatial structure of regional development and land exploitation, and government performance assessment and macro-control.

(6) Demonstration of the application of key technologies in the functional zoning of typical territory. This program will choose a number of typical functional zones in China and develop the simulation and evaluation technology for measuring the developmental effects of economy, society, resources and environment therein. Furthermore, it will draw up the standard, procedure and quantitative method for the acquisition and processing of different sequences of indicators, develop the technology for indicator integration and the overlay

and transformation of human-nature spatial unit, build up the intelligent, real-time methods and models for the comprehensive evaluation of different types of zone, develop the simulation and evaluation technology for measuring the implementation effects of different functional zoning schemes, and come up with technical specification for the formulation of spatial planning scheme for different types of territorial functional zones.

4.1.4 Key Scientific Issues and Innovations

Key scientific problems are: 1) the formation process and identification method of territorial functions in China; 2) calculation algorithm for the selection of indicators and the determination of spatial boundary in different types of functional zoning; 3) technologies for spatial information acquisition and processing as well as dynamic monitoring and evaluation of territorial functions; 4) development of spatial decision-making support system for the management of territorial function oriented zoning.

Major innovations are represented in two respects. Firstly, this program will theoretically and practically promote the development of the spatial equilibrium theory of regional development and the evolutionary theory of spatial structure, and achieve theoretical innovation through the combination of large-scale comprehensive investigations of regional development with the development of spatial equilibrium model. Secondly, this program will establish new classification system of territorial functions, put forward the method system and technological procedure for territorial function identification and functional zoning, achieve the formulation, integration and centralized management of the schemes and related technical documents of national and provincial territorial function oriented zoning, and improve the continuous service ability toward national spatial planning. Thirdly, this program will investigate the theories and methods of territorial function oriented zoning and the models and algorithm for information collection, processing, and spatial analysis, with an aim to lead the research frontier of methodology for territorial function oriented zoning.

4.2 Forecast of Urbanization and Overall Mechanism of Urban-Rural Integration

4.2.1 Significance of the Topic

Urbanization is a key international issue. In western developed countries, the existing studies of urbanization mainly focus on the process, mechanisms, effects and spatial governance of suburbanization, re-urbanization, and extended metropolitan area *etc*. Specifically, it includes five aspects. The first aspect involves the research on the processes and mechanisms of spatial agglomeration and diffusion in urban agglomerations. The second aspect is the research on the resource and environmental effects of the agglomeration of population and economic activity. The third aspect is concerned about the governance mechanism underlying coordinated urban development. The fourth aspect is to analyze the coupling relationships between industry and population on the one hand and land and environment on the other hand, and to simulate and forecast urban development trend by means of spatial analytical models such as GIS. The final aspect is to conduct research on the regulation of metropolitan areas. In the future, with the ever deterioration of environmental problems and the widespread acceptance of sustainable development concept, the smart management of urban growth and urban sustainable development will constitute the new focus of urbanization research[35,47].

Urbanization has always been a hot issue concerned by Chinese scholars and the integration of urban and rural areas is a key part of government decision-making vital to the achievement of modernization. In recent years, considerable progress has been made in rural labor force transfer, rural urbanization, development models of small cities and towns, regional differences in urbanization process, and the spatial agglomeration and diffusion in urban agglomerations, *etc*. However, further research on the driving mechanisms of urbanization at different stages, the regional pattern and evolving trends of urbanization, the resource and environmental effects of urbanization, and the mechanisms for the integration of urban and rural areas is still lacking. Since the reform and opening-up, many problems have emerged along with the continuous and rapid economic growth and large-scale urbanization. In China, the development of many cities has moved away from the normal track, and urban construction has tended to be chaotic or even out of control, which increasingly intensified the existing contradictions and conflicts between socio-economic development and resources, ecology, and environment, and brought about changes in the nature and characteristics of urban-rural conflicts. Such problems have posed significant challenges to the sustainable development of social economy in our country[3,48]. Therefore, in-depth study of future urbanization process, urbanization models, and governance mechanism for the integration of urban and rural areas in China is urgently needed in order to provide scientific basis and decision-making support for the formulation of healthy urbanization strategy and the coordinated development of urban and rural areas.

4.2.2 Research Objectives

According to the trend of chaotic urbanization and the new problems emergent in the process of coordinated development of urban and rural areas at present and based on the analyses of urbanization process in China, its driving mechanism and future development, future research should analyze the major territorial types of China's urbanization and their basic features, the key factors affecting urbanization process in China and their evolving trends, compare the resource and environmental effects of different spatial forms of urbanization, reveal the formation conditions and urban-rural integrative mechanisms of different urbanization modes through the case studies of such representative areas as Yangtze River Delta, Pearl River Delta, Urban Agglomeration of Wuhan, urbanized areas with decentralized distribution of cities and towns, and cities and towns in rural area, establish the geographic information system platform of representative cities and the dynamic simulation system of urbanization process at national level, and come up with policy recommendations for the regulation of urbanization and coordinated development of urban and rural areas.

4.2.3 Main Contents

(1) The process, driving mechanisms and future prospect of urbanization. We will learn from the urbanization experiences in other countries to summarize the relationship between urbanization and socio-economic development and identify the different phases of China's urbanization process and their respective features. In addition, we will analyze the driving mechanism and development models of urbanization process at different stages, carry out prospective analysis and forecast of urbanization process and spatial development strategy through multi-scheme modeling and scenario analysis.

(2) Research on the affecting factors and evolving trends of the regional pattern of urbanization process. Based on the comprehensive measurements of urban population, land use, and economic development in recent years, future research should attempt to systematically classify different types of regions in China's urbanization process, illustrate the basic characteristics of regional pattern changes in the urbanization process in China in terms of urban agglomeration, rapidly urbanized areas, analyze the key affecting factors and trends of the regional patterns of China's urbanization process, and forecast their evolving trends.

(3) Research on the resource and environmental effects of the spatial forms of urbanization. Based on the diagnosis and evaluation of major resource and environmental problems confronting China's urbanization, future research will explore the interactive mechanisms between the spatial form of urbanization and the restrictive factors of regional resources and environment, compare the resource and environmental effects of different spatial forms of urbanization, and propose the principles and criteria for the appropriate spatial forms of urbanization under the restriction of regional resources and environment.

(4) Research on the dynamic simulation of urbanization process in typical regions and the governance system to achieve coordinated development of rural and urban areas. In this research, a comprehensive investigation of representative cities in China will be conducted and the GIS and dynamic simulation system platform of typical cities will be established. Moreover, research will be conducted on the investigation of the forming conditions and urban-rural integration methods of different urbanization models through case

studies of such typical areas as Yangtze River Delta, Pearl River Delta, Urban Agglomeration of Wuhan, urbanized areas with decentralized distribution of cities and towns, and cities and towns in rural area.

(5) Research on the processes of urban-rural material and energy flow and the approaches to achieve coordinated development of urban and rural areas. This research will analyze the basic theories of urban-rural flows of material, energy, and other productive factors, evaluate the pattern of interest transfer amidst the process of urban-rural interaction, reveal urban-rural spatial-temporal differentiations, demonstrate the spatial-temporal trends of the future transformation of urban and rural areas, analyze the interactive relationships between industries and employment in urban and rural areas, investigate the basic approaches to achieve unified allocation of land resources in urban and rural areas, establish the operational system and governance regime for the coordinated development of urban and rural areas, analyze the development modes of urban agriculture and the regional models of China's new rural construction program, and explore the possible trajectories to achieve integrated development of "regional system of cities, towns and villages".

4.2.4 Key Scientific Issues and Innovations

The first key research question is to simulate urbanization process and develop comprehensive GIS and dynamic simulation platforms of typical cities. The second key research question is to analyze the driving mechanism of urbanization process and determine the parameter of different development scenarios. The third key research question is to model the spatial evolutionary process of population, industry, and construction land amidst urbanization process and to investigate the coupling relationships between such process and resource environment. The final key research question is to analyze the interactive mechanism between the spatial form of urbanization and regional resource and environment.

Main innovations include the following four aspects. The first aspect is to characterize the urbanization processes under different scenarios and conduct integrated study of the affecting factors of urbanization process from the perspective of coupling between pattern and process. The second aspect is to forecast the evolving trends of the key driving forces of ecological environment and urbanization process by means of integrated coupling model, and to measure urban development capacity and the environmental effects of urbanization in conformity with the dynamic changes of ecological environment in typical regions. The third aspect is to investigate appropriate urbanization modes based on the distinctive nature of China's urban development and the important restrictive objective of achieving coordinated development of urban and rural areas. The fourth aspect is to adapt urban dynamic evolutionary model to the Chinese context by introducing both controlling factors and restricting conditions, and to establish integrated simulation platform of the comprehensive system of urbanization by making use of GIS environment, massive spatial data management system, and computing tools.

4.3 Ecological Compensation Theory and Approaches to Achieve Coordinated Regional Development

4.3.1 Significance of the Topic

Entering the new century, China's regional development policies have gradually shifted from either balanced or non-balanced development to coordinated development, which refers to coordination in urban and rural development, regional development, economic and social development, human and nature development, as well as domestic development and opening up to the outside world[13,19]. Comprehensive, coordinated and sustainable development of nature, economy and society has replaced economic growth as the major objective and value orientation of regional development. But at the same time, regional disparity in China has widened continuously. In some areas, industrialization and urbanization are pushed forward irrationally regardless of the carrying capacity of resources and the environment, which has led to the ever deterioration of local and even regional eco-environment[3,37,49-51] .Against this background, it is necessary to promote industrialization and urbanization in a rational way and attempt to achieve the equalization of basic public services nationwide on the basis of regional carrying capacity of resources and environment, economic development potential and intensity in order to build up an orderly spatial structure of land development, which entails the establishment of comprehensive ecological compensation mechanism and policy support system[13].

Drawing up externality theory, public goods theory and ecological assets theory, the traditional theory of ecological compensation focused on passive protection and improvement of ecological environment, on the basis of the mechanism of self-organized evolution, feedback and restoration and the ethical foundation of sustainable development[52,53].

However, the development of social practice has accelerated the constant innovation of the traditional theory in several aspects. Specifically, coordinated development of nature, economy and society has replaced the traditional single target of eco-environmental restoration and protection and the combination of government regulation and market mechanism has replaced the traditional mechanism monopolized by government. In addition, the comprehensive standard including ecological benefits, social acceptability and economic feasibility has replaced the single standard of ecological benefits, while industry support and assistance in infrastructure construction have enriched the content and form of ecological compensation. Keeping informed of the research frontiers at home and abroad and building up the theoretical framework of ecological compensation with Chinese characteristics are not only a significant practical issue for solving China's realistic problems, but also an important way to promote theoretical innovation and discipline construction.

For a long time, government projects are the main form of ecological compensation in China with the goal of improving and protecting ecological environment[54-56]. However, there is still some way to go if we are to achieve the harmony of ecological, economic and social systems and to meet the requirement of regional equalization of basic public services in the new period. Many problems implicit in the existing ecological compensation mechanism have prevented the inhabitants in ecologically protected areas from sharing the fruits of reform and development and enjoying the same level of public services as their counterparts in economically developed areas. Thus, we should strengthen the research on ecological compensation mechanism, achieve the rational integration of regional development theory and ecological compensation theory, and construct a highly effective and rational regional eco-compensation mechanism featured by scientific design, multiple compensation objectives, explicit compensation standards, diversified compensation methods, long-term compensation period, reasonable policies and measures, and improved evaluation criteria[57].

4.3.2 Research Objectives

The research objectives are to coordinate the development relationships among different functional areas, achieve regional equalization of basic public services, develop the theoretical system and implementation approaches of regional ecological compensation, build up spatial-temporal GIS model for regional eco-compensation through the analysis of the patterns of ecological functional zones, investigate the ecological compensation schemes with diverse subjects, multiple methods, and at different spatial-temporal scales, put forward the assurance procedures and policy system for the implementation of ecological compensation; design the indicators for evaluating the implementation effects of ecological compensation, summarize the policy options to reconstruct new regional development relationships, and to scientifically evaluate the implementation effects of ecological compensation and its impact on the regional development pattern in China.

4.3.3 Main Contents

(1) Development of the theoretical system of ecological compensation. Based on the analysis of the background, point of departure and objectives of traditional ecological compensation theory, this research aims to test the relevance of such theory in the Chinese context, and develop the theoretical framework with Chinese characteristics through deconstruction and reconstruction of mainstream theories.

(2) Determination of the pattern of ecological functions in China. According to the function types and spatial patterns of different ecological systems, this research aims to identify the ecological function zones in China, classify them into different groups based on their importance and vulnerability

at different spatial-temporal scales, clarify the service areas of different levels of ecological functions zones, analyze the inter-regional relationships in the construction, maintenance and utilization of ecological functions, analyze the interactions among ecological function zones, clarify the subject and object of ecological compensation, illustrate the pattern of regional ecological compensation comprehensively and clearly, analyze the intensity of the interaction among ecological function zones, discuss the standards and methods for allocating inter-regional rights and interests according to the principles of ecological environment protection, social equity and economic feasibility, and develop a spatial-temporal GIS model for regional ecological compensation and present the scheme through visual expression.

(3) Implementation schemes of regional ecological compensation. This research attempts to comprehensively analyze the subject's compensation capacity and the object's willingness to accept compensation, design the implementation schemes of regional ecological compensation in China, analyze inter-regional interactions in ecology and economy, put forward the methods for the identification of both subject and object in ecological compensation, clarify the type and intensity of the cost-benefit relationships between the subject and object as well as the implementation schemes at the present stage, analyze the roles played by different subjects such as the government, market, non-governmental organizations, *etc.* in regional ecological compensation, investigate the possibility of various ecological compensation models, design the compensation models and their various combinations through detailed investigation of the characteristics of ecological compensation at international, national, and regional scales, evaluate the effectiveness of different compensation methods, such as financial assistance, industry support and infrastructure construction assistance, *etc.*, design diversified compensation models for different resource-based zones and ecological function zones by taking into account inter-generational differences, different stages of economic and social development and different time periods, and draw up correspondence measures in the aspects of policy, law, finance, industry, population transfer and so on to ensure the smooth implementation of regional ecological compensation schemes.

(4) Evaluation of the implementation effects of ecological compensation schemes in typical areas. Based on the selection of some representative areas, this research aims to establish the comprehensive evaluation system involving ecological, social and economic elements, assess the implementation effects of ecological compensation in such respects as ecological and environmental protection, the equalization of basic public services and the development of distinctive local economy, clarify the relationships and weightings of three indicators, *i.e.* ecological and environmental protection, economic development, and social harmony, establish a comprehensive and reasonable evaluation system and criteria, and conduct a comprehensive analysis of the impact of regional ecological compensation mechanism on China's future

eco-environmental protection and socio-economic development through description and systematic simulation.

(5) Research on inter-regional interest transfer mechanism. Drawing up market mechanism principle, product transaction concept and the existing literature on ecological compensation mechanism, this research is intended to investigate the design of Clean Energy Quota System and environmental capacity (Pollutant Emissions Right) trading system.

4.3.4 Key Scientific Issues and Innovations

Key research questions include: 1) the methods for identifying and classifying ecological functions in China; 2) the standard and calculation method of inter-regional allocation of rights and interests and the implementation schemes of inter-regional ecological compensation; 3) the development of theoretical framework of ecological compensation and the evaluation of the implementation effects of ecological compensation schemes under such framework.

The main innovations of this research are: 1) the preliminary development of the theoretical framework of ecological compensation which has satisfied national needs, promoted disciplinary construction and has filled in the gap between theory and practice; 2) the achievement of "win-win" objective between ecological protection and economic development through the combination of regional ecological compensation mechanism and coordinated regional development; 3) the achievement of dynamic monitoring and evaluation of the implementation effects of ecological compensation schemes and the design of related policy system and institutional arrangements to ensure the smooth implementation of the schemes.

5 Strategies and Actions of Chinese Academy of Sciences

The coping strategies of Chinese Academy of Sciences are to strengthen interdisciplinary research to meet the national needs in resource environment and sustainable development, adjust and optimize the institutional arrangements of Chinese Academy of Sciences in terms of performance evaluation, institute construction, and cross-disciplinary cooperation, accelerate the development of theories, methodology, and technical platform in regional development at the Chinese Academy of Sciences to enhance the global competitiveness of its regional development research team, and build up "Simulation and Decision Support System for Regional Sustainable Development in China" through the combination of long-term research plans and short-term research projects.

5.1 Basic Strategy and Overall Deployment

5.1.1 The Development of Regional Development Research as An Interdisciplinary Field Requires the Removal of Man-made Obstacles and the Employment of Multiple Means

Regional study is a comprehensive and cross-disciplinary field which requires long-term accumulation of knowledge in both natural and human sciences. As a result of socio-economic development, the role of science and technology in addressing a series of major issues confronting humankind is increasingly strengthened. In this process, the trend of cross-disciplinary fertilization is also becoming more pronounced. A large number of natural science research (findings) in resource and environmental science need to find an outlet in the field of sustainable development. However, Chinese Academy of Sciences has always focused exclusively on natural sciences and some leaders and scholars at the academy set up many barriers to the growth and development of cross-disciplinary research in practice while in theory they also acknowledge the importance of such research in the solving of major practical problems and the promotion of disciplinary construction.

In this sense, we need to extend the research direction of resource and environment to include regional development. Resources, environment, and regional development are actually three closely connected subfields. It will be difficult to figure out effective solutions to resource and environmental problems if we analyze them separately and independently. In fact, both science development and key human practices demand a breakthrough in philosophy to go beyond natural sciences, as most of the important scientific and socio-economic questions in contemporary era are cross-disciplinary and comprehensive and the answer to these questions will undoubtedly entail the cross-fertilization and cooperation of multiple disciplines, in particular between natural and social sciences.

Previous research on regional development at the Chinese Academy of Sciences is mainly concerned about natural science. Although we have established relatively strong competitive advantage in this field, we still have a lot of work to do in view of the huge amount of demands for regional studies and the development trend of other research institutes in the future.

Research on resource, environment, ecology and regional development is actually an integrated whole composed of natural foundation and socio-economic development. However, due to the setup of research divisions, real cooperation is difficult to achieve in actual research work.

5.1.2 Develop Theoretical System of the Interaction among the Elements of Man-earth System and the "Environmental-social Dynamics" of Man-earth Relationship

The determination of the strategy and research direction of "man-earth relationship" and regional development requires us to follow the principles of "man-earth areal system" and the correlation between resource environment and development in order to provide scientific basis and schemes for solving the major issues in territorial and regional development and to develop our distinctive theoretical system and methodology through the improvement of independent innovation capability. Such research should deal with fundamental, comprehensive and strategic issues and play an important role in the implementation of national development planning and the opinion of "Five Integration". The main contents of the research include the characteristics and evolutionary theory of man-earth areal system, the interactive mechanism between environment and development and among their sub-elements, *i.e.*, the "environmental-social dynamics" of man-earth relationship, the relationship between the process and pattern of regional development, the prediction of the coupling relationships between environment and development at different development stages in future China, and the application of integrated research and methods.

The earth surface system consists of two parts, namely, "earth" and "man". The former refers to natural elements, including water, soil and ecological environment. The latter, on the other hand, is the socio-economic aspect which includes population, economy, urban and rural areas, industry and agriculture, and transportation, *etc*. The internal mechanism of natural system is not our concern. Instead, the research will focus on the interactions between the two elements of "man" and "earth" on the earth surface from a regional perspective, and clarify the basis of regional development. The research on the characteristics and evolutionary theory of man-earth areal system is the prerequisite to achieve scientific planning and directed regulation of regional development in accordance with human needs. Thus, the characteristics of the system must be illustrated. Large amount of data and research findings have shown that man-earth areal system is an open or semi-open integrated system with the exchange of materials, energy and information with the outside world. "The idea of changing the world of humankind stimulates the interaction between human beings and environment, and human will modify their behaviors according to the results of such interactions, the variegated and complicated ways in which human change and adapt to environment create the kaleidoscopic world on earth. The world is an interdependent network system"[58]. The major method to understand this "interdependent network system" is to reveal its characteristics and conduct zoning by type.

The relationship between "man" and "earth" is in nature the same as that between society and natural basis. How does this relationship affect national and regional sustainable development? To answer this question, we need to analyze the interactive mechanism among the various elements of environment and development, *i.e.*, the "environmental-social dynamics" of man-earth relationship or alternatively called by American scholars as the mechanism of environmental-social dynamics. We need to make judgments and early warnings of the environmental and developmental status at different development stages of the nation and various regions through mechanism analysis, reveal the correlation between the process and the pattern of regional development, and establish our own theoretical system of urbanization and regional development.

How to analyze the "environmental-social dynamics" of man-earth relationship? The man-earth areal system confronting regional development researchers is an extremely complicated system. To understand this system, we must bring the methodological strength of geography into full play, and at the same time take advantage of the methods of system science, ecology, and economics, *etc*. We should develop integrated methodology based on these existing methods. Such integration includes the integration of historical data and information, the integration of elements and their functions, the integration of varieties of correlations among different regional development status, and so on. The application of the integrated methodology demands new innovations by scholars on the basis of traditional methods.

The main objectives of integrated research are the integration and coupling method of the territorial differentiations of natural and human elements. We cannot understand the characteristics of the territorial differentiations on earth surface and the sustainability of socio-economic development without such integration and coupling. We can make integrated (type) zoning, *i.e.* the natural-socio-economical-ecological zoning, or what might be termed ecological and economic zoning on that basis. The research on such zoning is of great theoretical and practical significance to the function orientation of future national and regional development.

5.1.3 Construct Data and Technical Support Platform, and Achieve the Prediction and Forecast of Regional Development Trends

For a long time, it has been difficult to accurately understand the affecting factors, process and evolutionary mechanism of regional development. It seems that anyone can put forward his opinions in this field. However, the success and failure of a large number of decision-making processes demonstrate the rich content of this field. Pushing forward the construction of data and technical support platform, establishing the basic database of regional development, developing and integrating the analytical and forecasting tools (models), and improving the ability of visual expression are all necessary means to enhance our regional development research capabilities. Therefore, we should emphasize the construction of key laboratories of this field at the Academy.

The construction of technical support platform for regional and urbanization development research is vital to the enhancement of our research capability and independent innovation capability in such subfields. To strengthen the application of data mining and other new technical means are not only the requirements of disciplinary construction, but also the international trend in this field. The integration of existing dispersed data resources is an important foundation for the improvement of both depth and visibility of urbanization and regional development research. During the long period of time to come, a large amount of research work is needed in urbanization and regional development to support national regulation and governance.

5.1.4 Establish Cross-disciplinary Joint Research Center and Achieve Cross-fertilization and Integration

The problems at current stage are prominent. The researches on resource, environment, ecology, and regional development is actually an integrated whole composed of natural foundation and socio-economic development. However, due to the setup of research divisions, real cooperation is difficult to achieve in actual research work.

The Chinese Academy of Sciences should establish a joint research center, drawing upon the inspiration from various disciplines of geography, resource science, and ecology. Such research center can be further enlarged to become part of central ministries and departments (such as the Development Research Center of the State Council, and the Academy of Macroeconomic Research of National Development and Reform Commission). The final objective is to establish research and forecasting center of regional development and territorial regulation.

5.1.5 The Construction of Research Institutes and Talents Training

The research team should be further expanded. In our Academy, there are less than 50 researchers working on regional development and most of them are affiliated with the Institute of Geographic Sciences and Natural Resources Research. This number is far from enough in view of the huge amount of national demands. Therefore, the research team needs to be expanded in order to strengthen our competitiveness in this field.

A long-existing problem is that only a few institutes established research divisions to conduct regional development research at Chinese Academy of Sciences. They include the Department of Human Geography and Regional Development (with four subordinate research groups) at the Institute of Geographic Sciences and Natural Resources, and other departments at Nanjing Institute of Geography and Limnology, Northeast Institute of Geography and Agroecology Ecology, and Xinjiang Institute of Ecology and Geography. Nowadays, there are few researchers of regional development at Chengdu Institute of Mountain Hazards and Environment. Due to inadequate understanding of cross-disciplinary research and the personnel evaluation criterion based on SCI, many excellent talents have to engage in administrative work in governmental departments at all levels or go abroad.

The above-mentioned research centers and laboratories of regional development needed to be expanded in the next 10 years, and efforts need to be made to establish new research institutions with regional development as the main research focus.

5.2 Implement Long-term Research Plan and Establish the Simulation and Decision Support System for Regional Sustainable Development in China

According to the orientation of Chinese Academy of Sciences, the main objective of regional development research is to play the role of "think tank". The research plan of "Simulation and Decision Support System for Regional Sustainable Development in China" will be carried out in the following 20 years to establish a leading research center of regional sustainable development in China, which can serve governmental policy-making and management, contribute to the public education of regional sustainable development concept, and facilitate the participation in international research network of sustainable development.

The objective of the research plan is (Fig. 5.1) to establish the Simulation and Decision Support System for Regional Sustainable Development in China at the Chinese Academy of Sciences in the next 20 years through spatial analysis, mechanism research, model building and application and scenario analysis and visual demonstration, drawing up regional sustainable development theories and theories from other disciplines such as earth science, system science, computing science and social science and supported by modern computer techniques and the observation and collection of data at multiple spatial scales and in various forms. The establishment of such system can help diagnose the status of China's regional sustainable development, forecast the developmental trend, simulate the policy effects, and provide a comprehensive platform of diagnosis, early warning, and visual expression for government, enterprises, and the public to better understand the objectives, status, prospect and decision-making process of regional sustainable development in China.

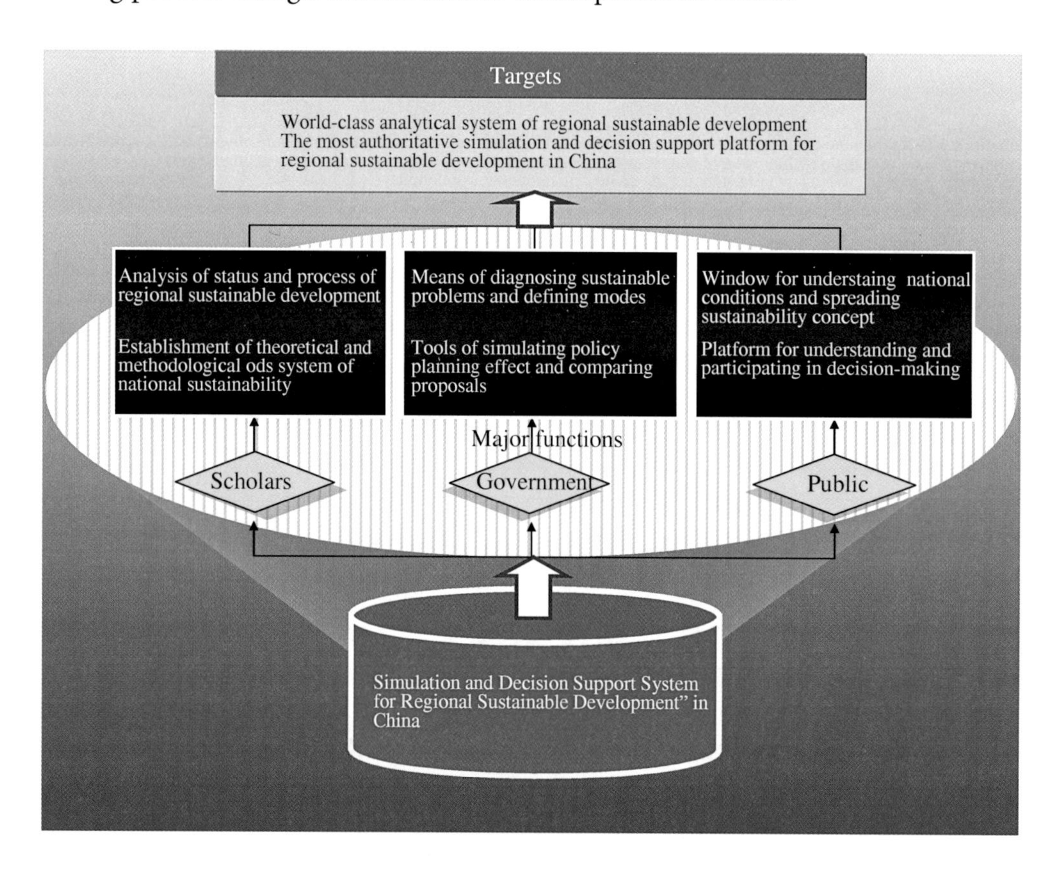

Fig. 5.1 Objectives and major functions of the "Simulation and Decision Support System for Regional Sustainable Development in China"

Advantages and Disadvantages of the Chinese Academy of Sciences in Regional Development Research

Advantages

(1) CAS has a comprehensive science system of "resources-environment-sustainable development" and a distinctive feature of cross-fertilization between natural and human sciences. Therefore, it can provide a solid support for the illustration of the nature of regional development and the solving of key problems in national and regional development.

(2) An influential regional development research team with reasonable age structure has formed after years of research and accumulation. The nature of the CAS has avoided the interest conflicts between departments and local governments. As a result, it has received more and more recognition from decision-makers and the public for the authenticity and fairness of its research findings. This is the advantageous condition for the further enhancement of the role of CAS as the "think tank" of national development strategies.

(3) CAS has formed a reasonable pattern of the spatial distribution of its affiliated research institutes around the country which is consistent with national and local development needs and the fundamental objective of disciplinary construction. There is an urgent need for us to further integrate research resources, improve our independent innovation capability, and solve significant problems through the joint efforts of multiple disciplines.

(4) The long-cultivated culture of innovation, cooperation and dedication of regional development scholars at our Academy are also important psychological and cultural resources. Moreover, a great deal of experience has been accumulated in organizing and managing research plans and program implementation through years of operation of scientific research plans and key programs.

Disadvantages: There is a shortage of research capability in social sciences and there exists no such research directions as institution, culture and behavior. Social and economic research mainly depends on statistical data without necessary means and tools for data collection. Long absence of significant research plans leads to the lack of cooperation and integration among multiple disciplines. Resource sharing has just begun, and there exists a sharp contrast between research visibility and research capability. There is also a considerable gap between our research outcome and national strategic demands.

The key scientific questions of the research plan include: What are the objectives and models (standard curve of regional sustainable development) of sustainable development in different regions? How to obtain the data to describe the process and status of regional sustainable development? What are the threshold values of early warning for the "Red Light Region" and "Yellow

Light Region" in regional sustainable development? What are the mechanisms and optimization approaches of regional sustainable development under the restriction of multiple objectives of saving resources, protecting environment, improving livelihood and enhancing competitiveness?

Based on the above questions, the research objectives are to evaluate the sustainable development conditions (abilities) of different regions in China, put forward sustainable development objectives and models for different regions, develop different indicator systems for the sustainable development of different regions, establish corresponding data collecting and processing system, dynamic monitor regional sustainable development processes, build up the evaluation model of regional sustainable development, analyze the status of regional sustainable development, forecast the trends of sustainable development; establish the decision support system for regional sustainable development based on man-machine interaction, scenario analysis and visual expression, regulate and optimize the process of sustainable development, and simulate the prospect of regional sustainable development. (Fig. 5.2)

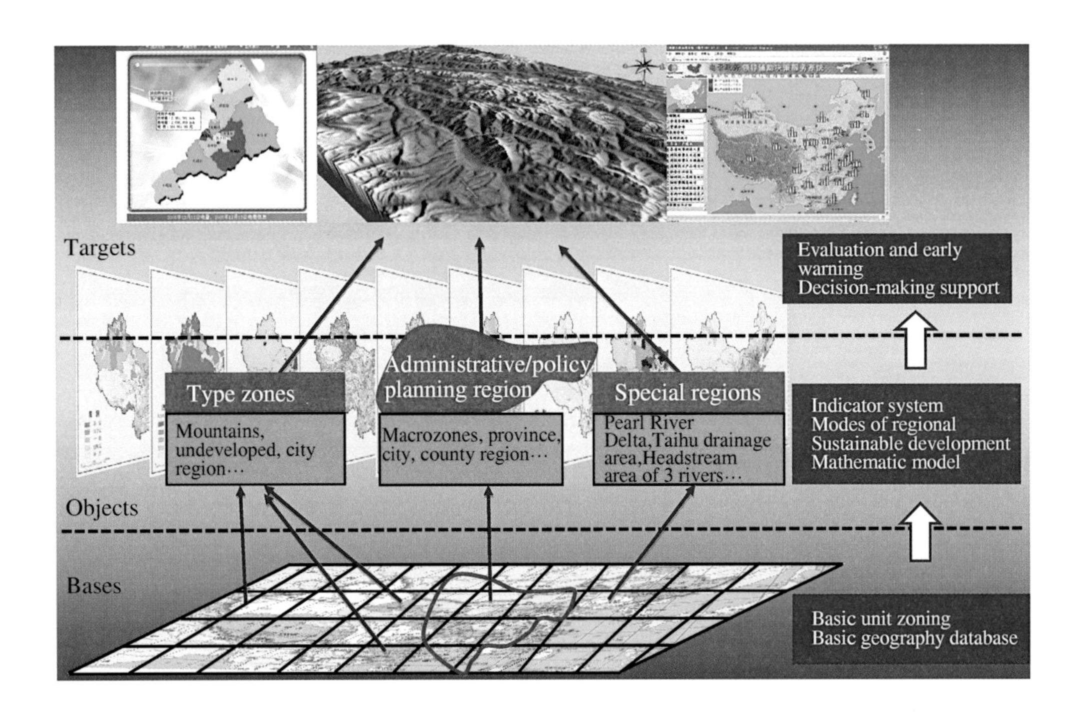

Fig. 5.2 Basic structure of the "Simulation and Decision Support System for Regional Sustainable Development in China"

The technical scheme of this research includes the following steps. The first step is to delineate the basic units in order to establish the basic geographical database in conformity with changing regional division scheme. At this step, a catalog of basic geographic information should be created and basic geographic data should be collected by various means (such as statistic data, field station data, remote sensing data, field survey data, *etc.*) to establish data collection network and basic database. The second step is to conduct research on the theories and mechanisms of sustainable development in different types of regions, work out the evaluation indicator system of sustainable development, and develop both general and specific models of regional sustainable development. Research areas cover broadly to include mountainous region and urban agglomerations classified by sustainable development type, eastern coast classified by administrative division, and special regions such as after-Earthquake Wenchuan, Taihu Lake drainage area classified according to national or local development needs. The third step is to carry out evaluation and early warning of the sustainable development status in different regions, and provide decision-making support for and visual expression of regional sustainable development.

Major achievements include (Fig. 5.3) the completion of the "database and model base" of China's regional sustainable development, the promulgation of evaluation and early warning reports on China's regional sustainable development, the gradual development of the theoretical and methodological system of sustainable development, and the formation of a research team of "Simulation and Decision Support System for Regional Sustainable Development in China" with international influence which is composed mainly of geoscience scholars at Chinese Academy of Sciences and has been well recognized by Chinese governments and the society, the establishment of the simulation and demonstration platform of the processes and scenarios of regional sustainable development and the visual expression platform of the decision-making process and its effects.

As for the organizational pattern (Fig. 5.3), CAS should strengthen cross-disciplinary study, gradually form a stable research team based on the research base of economic geography and the joint efforts of physical geography, resource science, ecological science, environmental science, computing science, system science, and social sciences, *etc*., build up a comprehensive technique support system for dynamic monitoring of regional sustainable development by making use of the field stations of ecological network and remote sensing techniques, construct database collecting and managing system, forecasting and early-warning model, and decision-making support system step by step, combine the existing research on development direction and models with the evaluation and early-warning of the development processes to promote the implementation of the research plan through "top-level design, integrated deployment, target management, separate implementation", and explore the synergy of CAS on the study of both general and locally specific questions of regional sustainable development through the joint efforts of research institutes inside and outside Beijing.

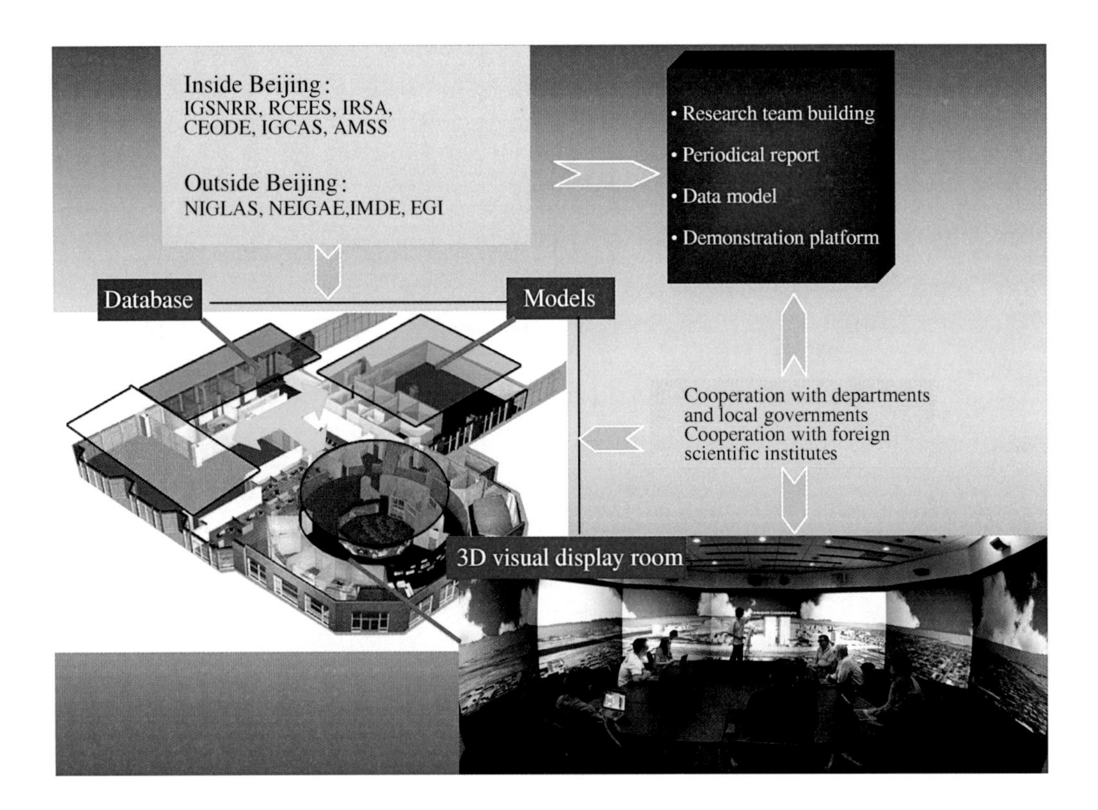

Fig. 5.3 The achievements and Organizational structure of the "Simulation and Decision Support System for Regional Sustainable Development in China"

5.3 Start Pilot-Research Programs and Establish Platform of "Simulation and Decision Support System for Regional Sustainable Development in China"

The construction of the platform of "Simulation and Decision Support Platform for Regional Sustainable Development in China" (platform for shorthand) is the main task of the long-term research plan of "Simulation and Decision Support System for Regional Sustainable Development in China", and an important basic infrastructure to improve the research capability of CAS and facilitate the implementation of China's sustainable development policy. The platform involves a relatively stable research team in the field of regional sustainable development consisting of researchers from a variety of disciplines and departments. It is an open research network with extensive international communication and cooperation that involves different departments and regions at various spatial scales. In order to be included in the research plan of "Simulation and Decision Support System for Regional Sustainable Development in China" in the 12th Five-Year Plan, provide scientific services for the society and decision-making support for the formulation of central and local 12th Five-Year Plan, and facilitate the implementation of the strategies and objectives laid out in "Roadmap of the Development of Science and Technology in Regional Development research during the additional years to 2050", the pilot-research programs of the research plan entitled "Basic Framework of the Simulation and Decision Support System for Regional Sustainable Development in China and its application in typical regions" should be kicked off as soon as possible.

The objectives of the pilot-research programs are to judge the basic status and future trend of regional sustainable development, clarify the basic framework of the platform of "Simulation and Decision Support System for Regional Sustainable Development in China", put forward the basic region types in the simulation and forecasting of sustainable development, solve the fundamental issues of research background, framework and regional differentiations underlying the implementation of the research plan of "Simulation and Decision Support System for Regional Sustainable Development in China", choose typical regions with long-term research foundation to test the technical procedure of the research plan of "Simulation and Decision Support System for Regional Sustainable Development in China" according to the requirements of pushing forward "Major Function Oriented Zoning", and provide solid basis for the implementation of the research plan of "Simulation and Decision Support System for Regional Sustainable Development in China" through the investigation of such key issues as sustainable development models, data collection, model construction, decision-making support, and application and demonstration that are closely related to the construction of platform.

Major tasks include:

(1) Basic status and the trend of regional sustainable development in China. This research should comprehensively assess the basic status of regional sustainable development in China, analyze the major affecting factors of the status, patterns and processes of China's regional sustainable development, and illustrate the formation mechanism of different models defined by different status and development processes, select different types of countries to systematically examine their experiences and lessons in the construction of regional sustainable development model and early-warning system, analyze the models and policy systems of sustainable development in different types of regions and reveal the trend of the strategic needs confronting China's regional sustainable development through in-depth case studies of different types of regions.

(2) Basic framework of the "Simulation and Decision Support Platform for Regional Sustainable Development in China", The research in this regard aims to analyze the major demands of the early-warning system of regional sustainable development, clarify the basic functions of the system, determine the major function modules of the system, clarify the major function of each module in the system, analyze their interrelationships, illustrate the operating

mechanism of the system, explore the major demands for the management and implementation of the system, determine the function modules supporting the management and implementation of the system, establish the structural framework for the management and maintenance of the system, and analyze the organization and management structure of the platform according to the related policies of state spatial governance.

(3) Basic types of regions in the simulation and decision support system for regional sustainable development. This research aims to analyze regional differences in the basic conditions of sustainable development, illustrate regional differences in sustainable development objectives and capability based on regional functions in national territory, summarize the schemes of regional divisions under different macro policies since 1949, classify regions based on different development objectives and models and in conformity with the needs of macroeconomic orientation and orderly spatial development, explore the criterion of data collecting in relevant fields and by functionally important departments, establish standard regions based on different objectives, models and patterns for the early-warning and evaluation of sustainable development ability through the analogy of different types of regions and in-depth study of typical regions, and explore the methods to transfer regional elements among different spatial scales.

(4) Application in typical regions. This research is to select typical regions with long-term research foundation under the guidance of "Major Function Oriented Zoning"to test the technical procedure of the research plan of "Simulation and Decision Support System for Regional Sustainable Development in China". It includes the models of sustainable development in typical regions, ways to obtain the data characterizing the process and status of regional sustainable development, threshold value of the early-warning of sustainable development, the mechanisms and optimization methods for the governance of regional sustainable development. Meanwhile, it should pay special attention to such key issues as sustainable development models, data collection, model construction, decision-making support and application and demonstration that are all related to the construction of platform construction.

Appendix 1

Monographic Studies on the Roadmap of Scientific and Technological Development in Regional Development

A1.1 New Factors and Mechanisms Affecting Regional Development

Rapid economic growth and the application of high technologies have dramatically reshaped the ecological, environmental and socio-economic structures in China, which in turn result in changes in the backgrounds and driving forces of regional development. The interaction between rapid economic growth and fluctuating socio-economic and environmental structures brings about new factors and mechanisms affecting regional development in China in different temporal and spatial contexts. One of the significant challenges that regional development researchers in China faces is to understand the regional effects and trend of these new factors and mechanisms, which is a prerequisite to the design of regional development roadmap.

A1.1.1 Literature Review on Factors and Mechanisms of Regional Development

The earlier researches on factors affecting regional development mainly concern various locational factors underlying industrial distribution, as represented by the efforts made by such distinguished scholars as Thunen, Webber, Losch, Christaller *etc.*[59,60] Since the quantitative revolution in economic geography in 1950s and 1960s, western scholars have paid considerable attention to economic factors and mechanisms including capital accumulation, technological improvement, import and export trade, spatial agglomeration economy, industrial linkages, and technological gradient and so on. The "institutional turn" and "cultural turn" in new economic geography has emphasized the importance of non-physical and even non-economic factors, which further weakened the significance of physical factors on regional development[61,62]. During the past two decades, severe problems of global climate change and sustainable development have been arousing the awareness of regional researchers on the impact of physical factors on regional development[63]. A series of complex problems such as global climate change, global change of water recycling, land use and land coverage change as well as regional and global resource utilization have attracted more and more attention from regional development scholars who call for a comprehensive regional development research in which both physical and human factors should be taken into account[64].

After years of practices, regional development researchers in China have gradually realized that regional development is affected and restricted by multiple factors at different temporal-spatial scales including location, natural environment, resource endowment, state policy, historical foundation, globalizing economy, industrial transfer *etc.*[14,65,66] During the past decade, they have increasingly emphasized the interconnection and integration of physical and human factors under the philosophical guidance of man-earth areal system, in which regional development is considered as a non-linear synthesis of economic, social, ecological and environmental processes[67,68].

Generally speaking, the theoretical and methodological focus of regional development research varies in different countries due to the differences in development phases and geographical environments. As time goes on, factors that should be taken into account in regional research become more complicated, from traditional factors such as transportation, mine, water, to the increasingly important elements in recent years including globalization, technology and knowledge innovation system, and culture. All in all, research on factors underlying regional development in China is characterized by a gradual shift from single-factor analysis to integrated and synthesized multi-factor research.

A1.1.2 Scientific and Technological Demands for Researches on the Factors and Mechanisms of Regional Development

A number of factors such as resource security, global climate change, environmental regulation, the restructuring global financial system, reform of administrative district, the aging population, and regional culture will possibly have significant impact upon the regional distribution of socio-economic development and industrial structures in China in the future, under the control of the macro-pattern of physical environment[69-74]. Influenced by these factors, metropolitan regions will continue to be the major spatial organizations leading Chinese regional economy, and border economic cooperation zones have great potentials to become new growth poles of regional economy. In addition, the inter-regional division of labor and cross-regional collaboration will change from horizontal toward vertical form. Regional economic gap between eastern and western areas will probably be widened, while the gap in terms of income per capita is expected to be narrowed. In general, the analysis of these factors and mechanisms constitute the scientific and technological demands in regional development research as follows.

(A) Research on the Guarantee Capacity of Regional Resource Security

The strategic status of natural resources has not been lowered with the rapid development of science and technology. On the contrary, resource security has become an important element of national security strategy. It is necessary to study the guarantee capacity of resource security in strategically important regions especially in China where significant regional disparities in resource endowments and socio-economic conditions exist. Regional resource security refers to the status and capability of a certain region to economically ensure sustainable, stable and timely supply of sufficient natural resources. The realization of regional resource security in China would be able to guarantee a region to acquire enough resources at appropriate costs to support the sustainable development of its leading, related and basic industries at the present time and in the future[75]. In this point, several topics need to be investigated, including the factors that influence guarantee capacity of regional resource security, the evaluation indicator system of the status of regional resource security, the guarantee capacity of regional resource security under different scenarios of socio-economic growth and geopolitical conditions.

(B) Externality-based Inter-regional Interests Coordination and Compensation Mechanisms

Inter-regional relations have always been complex in China, especially under the background in which regional division of labor is gradually deepened, such issues as resource allocation (*e.g.* south-north water diversion, west-east gas transmission, west-east electricity transmission), ecological construction (*e.g.* the governance of water and land, the regulation of drainage area) and responsibility division (*e.g.* carbon emission reduction)[76] have further intensified this complexity. In an open environment, any changes of economic development and ecological environment in a region will inevitably influence the development of other regions or even the entire external system[49]. The existence of such inter-regional spillover effects requires concerned regional actors to share the benefitable incomes produced by positive external effects, at the same time to assume the costs by the negative external effects. In this sense, the study on the externality-based coordination and compensation mechanisms of inter-regional interests would be a long-term topic in the field of regional development in the coming future. Several subjects need to be investigated including game analysis on the regional stakeholder, the methods of defining the responsibility rights, and interests amongst regions, the mechanisms, models and regulatory measures of inter-regional compensation *etc*.

(C) Pre-warning Study on Regional Industrial Structures Under the Constraints of Sustainable Development

The upgrade and transformation of industrial structure reflects the phase of regional economic development and the condition of sustainable development. Study of the status and climate index of regional industrial structure under resource and environmental constraints is the precondition to establish harmonious inter-regional relations and achieve long-term sustainable economic development. Because industrial structure transformation is at the heart of regional sustainable development, it is imperative to construct a series of pre-warning methods, technologies, models and signal system in order to monitor the association degree between industrial structure and the base and supporting system of regional resources and environment, establish the climate index of regional industrial structure and to assess the risk degree or intensity of regional industrial structure imbalance[77]. Topics in this respect include how to develop the pre-warning indicator system for monitoring regional industrial structure, how to divide the warning range of the regional industrial structure and how to assess the contribution value of pre-warning indicators.

(D) Research on the Regional Dynamics and Spatial Pattern of New Economy Industry Development

Human society is now experiencing a significant transformation from industrial era towards informational one with profound changes in industrial types and the underlying strategic resources[78]. Although China as a whole is still in the middle stage of the industrialization process, knowledge and information-intensive industries have played a vital role in several developed cities in the eastern coastal area and developed more rapidly than traditional industries. The extent of dependence of these new industries on regional factors is quite different from that of traditional industries, which render several classical theories based on the distance decay-based rule to be ineffective. New economic activities such as e-commerce, virtual economy and logistics economy display new forms of spatial organization, which are different from those of traditional industries. Spatial redistribution and reorganization of different types of new economic activities will possibly produce new regional nodes. Analysis on the spatial features and regional dynamics of these new industries is helpful for our understanding of the spatial pattern of future regional development in China.

A1.1.3 Overall Goals and Stage Objectives to 2050

The overall objective of regional development research on factors and mechanisms is to examine the interaction between new and traditional factors that affect regional sustainable development at both intra-regional and inter-regional levels and their impact on regional resource security and the transformation of industrial structure. Specifically, efforts should be made to analyze traditional factors such as resource and environmental system, infrastructure system and new factors such as information, technological innovation, institution and culture. Focuses should be placed on the mutual interaction between the new and traditional factors and their effects on the evolution of regional industrial structure. Moreover, attention should also be paid to the pre-warning system of the sustainable development of regional industrial structure and theoretical and methodological issues of the coordination and compensation of inter-regional interests.

The major objective of regional development research on factors and mechanisms to 2020 is to analyze the spatial dynamics of the growth in new industrial sectors. It needs to study the spatial features and spatial process of the formation, growth and evolution of new industrial sectors, and to analyze the dependence degree of the new sectors upon regional economic, technological and institutional environment.

The major objective of regional development research on factors and mechanisms from 2020 to 2030 is to optimize regional industrial structure. Its main contents are to review the existing theories and methods concerning regional industrial transformation, to identify the factors affecting the process and direction of regional industrial transformation, to construct indicator system for measuring the evolution process of regional industrial structure, and to provide risk evaluation on the sustainable development condition of regional industrial structure.

The major objective of regional development research on factors and mechanisms from 2030 to 2050 is to focus on important theoretical and methodological issues, mainly including the study on cross-regional interest coordination and interest compensation mechanisms towards harmonious regional development and geographical equalization of basic public services, and to map the ensuring measures and policies of inter-regional compensation concerning multiple subjects, various modes and different temporal-spatial institutions.

A1.1.4 Implementation Plans and Technological Roadmap

(A) Major Scientific Issues

Based on the goals discussed above, the regional development research on factors and mechanisms in the coming decades mainly concerns the following scientific issues.

a. Research on the Affecting Factors and Assessment Index System of Regional Resource Security

Related topics are to analyze the impact and working mechanism of factors on regional resource security such as regional resource endowments, geo-political pattern, transport accessibility, socio-economic demands, technological improvement and resource substitution, *etc*., and to develop region-specific index system for evaluating resource security, and to assess the guarantee capacity of regional resources to support regional economic and social development, and to explore applicable regional resource security system.

b. Study of the Evolutionary Process and Optimization Model of Regional Industrial Structure under the Constraints of Sustainable Development

Based on the objectives of achieving the coordinated development of regional economic growth and social equality, of the carrying capacity of resource and environment, this study focuses on the analysis of the coupling relationship and relation coefficient among regional economic aggregate, income level, consumption structure, technological innovation capacity, employment structure and the carrying capacity of resource and environment. In addition, it also concerns the measuring of the coordination degree between industrial structure evolution, regional resource and environment basis, and technological and economic system; and the simulation of the evolution process of regional industrial structure and its social and ecological effects under different development situations.

c. Research on the Regional Dependence of New Industrial Sectors

The aims of this point are to identify mechanisms through which new factors such as informatization, regional culture and knowledge innovation system affect the locational choices and spatial organization of new industrial sectors including e-commerce, virtual economy, logistics and cultural industries, and to investigate the dependence degree of various new industrial sectors upon regional economic, technological and institutional environments, and to examine spatial relationships between new and traditional industries under the combined influence of new and traditional factors, and to explore countermeasures of regional development under different resource and environmental backgrounds and various technological economic conditions.

d. Study on the Function-oriented Inter-regional Interests Compensation Mechanisms

Based on the detailed analysis on the interest demands and conflicts among different stakeholders in regional development, this point is to investigate the logical relationship and allocation principles of inter-regional interests, and to discuss major-function oriented compensation models, methods and principles, and to explore corresponding supporting systems and ensuring measures.

(B) Research Design

Regional development research on factors and mechanisms is conducted at intra-regional and inter-regional scales according to the logics "analysis from new factor to comprehensive study on new and old factors, and to the pre-warning of regional industrial structure, and to regional compensation mechanisms".

Firstly, to conduct integrated comprehensive analysis on the influences and working mechanisms of new factors on regional development, for example geopolitics, global climate change and environmental regulation, aging population, social culture, informatization, *etc*.

Secondly, to study the guarantee capacity of regional resource security under the influence of both new and old factors. The focuses are to design the index system and to analyze the evolutionary process of regional resource supply and demand under different resource endowments and socio-economic conditions, and to construct regional resource supply and security ensuring systems that meet the requirements of regional competitiveness.

Thirdly, to construct the index system and pre-warning models by introducing the basic methods of economics and environmental sciences into the study on regional industrial structure, and develop the index system and pre-warning models of regional industrial structure under the constraints of sustainable development. The focuses are to identify the parameters, delimit threshold and simulate accumulation and to build the capabilities to forecast and predict the development trend, potential risks, and sustainable conditions of regional industrial structure.

Fourthly, to sep up a set of comprehensive and feasible methods and approaches concerning regional compensation mechanism, the determination of stakeholders' interests, compensation principles and financing in order to achieve the harmonious regional development in China.

Following to the above logics, two research projects can be set up at two stages.

Project 1: Research on the evolution trend and pre-warning models of regional industrial structure under the constraints of sustainable development

Main contents are:

- to study spatial dynamics of new industrial sectors;
- to study the spatial features and spatial process of the formation, growth and evolution of new industrial sectors, and to analyze the degree of reliance of different types of new sectors upon regional economic, technological and institutional environments;
- to analyze the evolutionary process of industrial structure and the affecting factors of this process;
- to identify the manner in which new and old factors affects the evolution of industrial structure and the interactive mechanisms between them, and to construct an indicator system for measuring the evolution of regional industrial structure;
- system dynamics modeling the evolutionary process of regional industrial structure under the constraints of sustainable development;
- to classify the various affecting factors of regional industrial transformation and to establish status analysis models under the analytical framework of stock, structure and flow;
- to explore risk and impact of industrial transformation;
- to identify the various risks associated with regional structural transformation, and to discriminate their various forms and the possible socio-economic problems they may induce, and to forecast their ways, extents, spatial scope and working mechanisms by using mathematical and statistical modeling;
- to research risk pre-warning mechanisms of regional structural transformation;
- to determine the value-at-risk underlying structural transformation based on mathematical and statistical modeling, which focuses on the

key thresholds and main parameters of abrupt and qualitative changes, to analyze the occurrence, accumulation, formation, evolution mechanism and major laws of the risks in industrial structures, and to explore the possible methods to prevent and control these risks.

Project 2: Research on the comprehensive accounting methods and compensation mechanisms of regional development

- to construct the theoretical system for inter-regional interests coordination and compensation mechanism;
- to establish theoretical framework in accordance with the regional characteristics of China based on the combination of existing theories of ecological compensation, regional division of labor, regional balanced development theory, and regional regulation theory;
- to analyze cross-regional interests conflicts and the negotiating process;
- to analyze the mutual interdependent relationships among the stakeholders in different regions in the aspects of economic, ecological and environmental development; to establish discriminating methods of the subjects and objects of interests, to clarify the cost-benefit relations, types and intensity between the subjects and objects of interests, and to investigate the complex game process of interests coordination and compensation;
- to evaluate the practical effects of inter-regional interests coordination and compensation effects;
- to analyze the compensation features of different types of regions and design corresponding compensation models; to explore the optimal combined models; to examine the effects of different compensation methods such as financial aids, industrial supports and infrastructure construction support; to design compensation effects in different temporal-spatial contexts of socio-economic development; to select representative regions to measure the real effects of inter-regional compensation.

(C) Roadmap Design

According to the scientific issues and research design illustrated above, the Roadmap of regional development research on factors and mechanisms can be constructed as follows (App. Fig. 1.1).

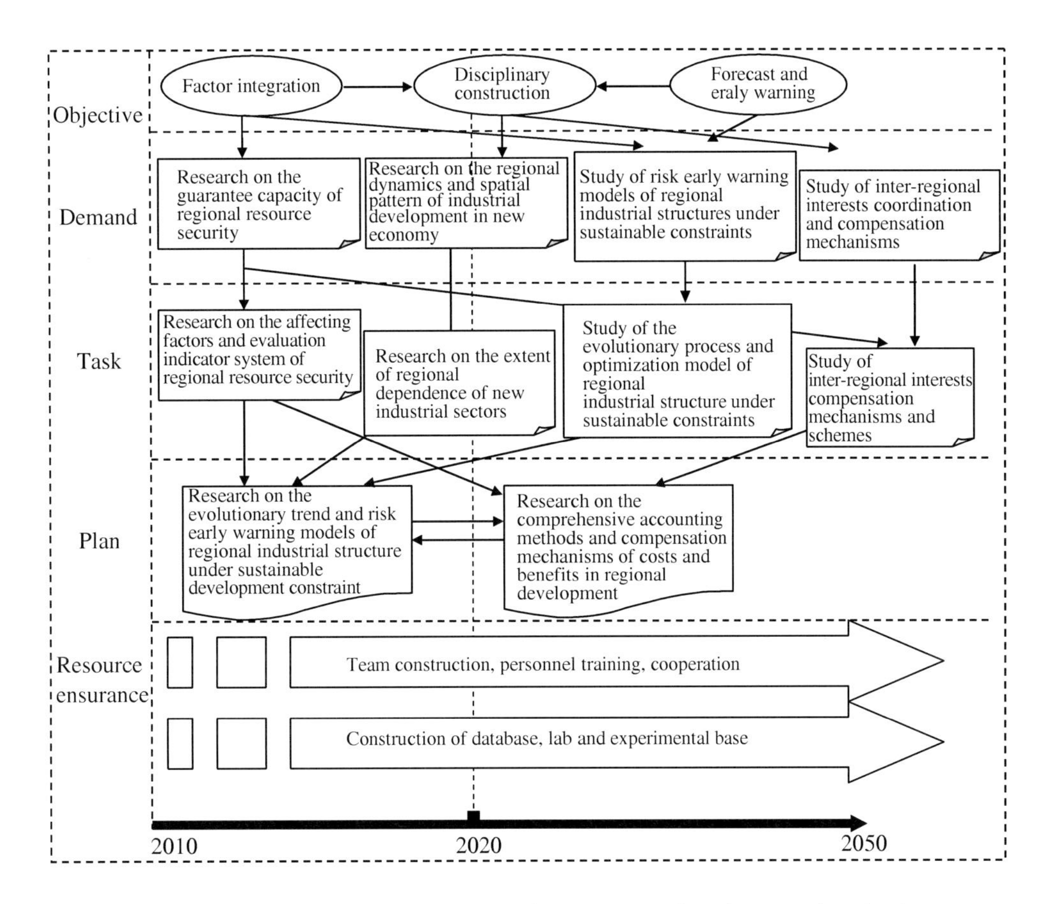

App. Fig. 1.1 Roadmap of regional development research on factors and mechanisms

A1.2 Roadmap of Spatial Organization of Industries

After more than 30 years of reform and opening-up, China has stepped into the middle stage of industrialization. And industrialization will still be the hot topic in the China's economic development within 20–30 years. However, the path to industrialization will be changed substantially under scientific outlook on development. Factors including resource, environment, globalization, informatization, servicization, *etc.* will exert profound impact on the trajectory and spatial organization of Chinese industrialization. Hence, it is necessary to construct the theories related to spatial organization in accordance to the characteristics of Chinese industrial development, by which the healthy and orderly development of Chinese industries can be foreseen and directed.

A1.2.1 Status Quo and Research Progress of Spatial Organization of Industries

Researches on the industrial development and spatial organization in China can be tracked back to the 1950s which mainly focused on the spatial scale and distribution of Chinese industrial development and the investigation on the industrial distribution of main regions from the perspectives of resource, transportation and national defense security[79-81]. During the 1970s and 1980s, the researches on Chinese industrial distribution were mainly carried out within the framework of regional planning, and involved resource, transportation and plant land-use, and regional water and soil resources, urban development and quantitative economic analysis as well[82,83]. In the 1980s, researches on Chinese industrial distribution were extended gradually from resources assessment and distribution of key enterprises to the systematic analysis of the principles of industrial distribution, relative conditions and affecting factors, and to the macro-development strategies[84,85]. At the same time, many investigations and researches were also conducted on the key issues such as techno-economic verification of industrial distribution, concentration and decentralization, regional industrial system, industrial regional complex, industrial location selection, and regional industrial structure, *etc.*[86-88] From different temporal and spatial perspectives, these researches have systematically summarized and developed the theories of traditional affecting elements of Chinese industrial distribution and spatial organization, industrial regional complexes and industrial bases, *etc.*, and have put forward a series of conclusions and modes, such as "pole-axis" diffusion theory of industrial space evolution, "T-type" organization system, *etc.*[89,90]

Since the end of the 20th century, the research focuses of industrial development and spatial organization have been shifted to the resource and environmental impact of industrial development, the formation mechanism of industrial agglomeration areas, locational changes of transnational corporations (TNCs), and new factors in industrial development[91-93]. These researches have provided important theoretical guidance for the optimization and upgrade of industrial structure in different regions, firm location and the enhancement of industrial competitiveness.

In recent years, with the emergence and rapid growth of new industries, the enhancement of the status of modern service industry in regional economy, the differentiation of the forms and the connection contents of industrial agglomeration, and the drastic changes in spatial organization pattern and evolution process of industries all over the world, a new era for the researches has been come on Chinese industrial distribution and the patterns and contents of spatial organization. Also, the spatial location of production elements has been changed thank to the progress of transportation, tele-communication and production technologies. Furthermore, the expanding global trade and the increasing FDI have resulted in changes in the spatial distribution patterns and organization of industries. The traditional boundaries of nations and regions were broken. The optimal location of industrial distribution is selected across world, the production sectors can purchase raw material and parts from all over the world, and a global market was formed. At the same time, some new spatial patterns of industrial organization appeared, such as TNCs, global production network and new industrial districts, *etc*.[94–98]

So far as China is concerned, the spatial pattern of industrial organization will also be changed in the future. The traditional extensive growth mode will be restrained, the level and tendency of the industrial concentration and geographical concentration of industries will be enhanced, and the pattern of cross-regional division of labor will be changed too. Specifically, the Middle and Western China will be the important industry bases of energy and raw-materials. The manufacturing industries of normal consumer goods will be gradually shifted to the inner-land and surrounding countries. The base distribution of equipment manufacturing industry will become more obvious. Electronic information manufacturing industry will still be agglomerated in metropolitan regions. Location-specific industrial bases and clusters will be formed around China.

A1.2.2 Research Trends and Scientific and Technologic Demands of Spatial Organization of Industries

In the coming 20–30 years, the speed and quality of China's industrialization process will be increased. However, industrialization will still be under the constraint of resource and environment and be influenced by uncertain factors of international economic and political situation.

(A) Industrial distribution under the resource and environment restriction

China's rapidly increasing economy has promoted the growth of energy and raw material industries, and greatly enhanced the production scale of ferrous metals, nonferrous metals, chemical industry and building material industry, which has thereby stimulated the demands for resource products, such as energy and raw-materials, *etc*. However, China is a country with relatively poor carrying capacity of population, resource and environment, which has led to China's increasing dependence on imported resources. In 2006, the amount of imported raw oil accounted for 44% of the total oil consumption, the imported ironstone accounted for 50% of the domestic demand, and the percentage for aluminum stone and copper ore were 33% and 50% respectively[99]. Besides, there are also huge regional differences in the carrying capacity of resource and environment. Therefore, it will be an important question to be discussed in the coming 20 years that how to adjust the regional industrial structure and optimize spatial distribution according to the regional resource and environmental restriction. It is necessary to analyze the impact of industrial activities on resource and environment on the one hand, and study the restriction of resource and environment on industrial activities on the other hand, which mainly include the ways and degree of the influence of major resource factors (such as water, soil and energy) on China's industrial distribution, and the ways and degree of the restriction of eco-environment conservation on China's industrial distribution.

(B) Spatial Organization of Industries Under the International Political Pattern

The increasing demand for energy and raw-materials driven by the rapid industrialization process in China has led to the rise in international energy and raw-material prices. And the price rise of energy and resources in turn directly or indirectly results in the increase in the production cost of iron and steel enterprises and the downstream industries, such as automobile, electrical household appliances, aviation and building material industries. In addition, the degree of China's export dependence was increased from 5% at the very beginning of reform and opening up to the 37% in 2008. The continuous and large-scale investment on iron-steel, textile and garment, common processing and assembling, light industries, non-ferrous metal, and building material industries has led to the rapid enlargement of industrial production scale. Actually, the excessive production capacity made them to be very dependent on foreign demand and international market. In 2008, the global financial crisis arisen by the US sub-loan crisis led to the slowdown of global economy, which directly affected Chinese exports of toys and electrical household appliances. As a result, many enterprises in the coastal area also closed down.

Therefore, confronted with the high level of dependence on international market, it is of great significance for the practical and theoretical cycles to probe into the impact of the uncertain international geo-politics and geo-economy on spatial organization of industries in China. The main contents include in what matter and to what degree Chinese industrial production depend on international market, the impact of different supply patterns of energy and raw material on Chinese industrial organization, and the status of China in global production network.

(C) The mechanism and spatial effect of industrial agglomeration and diffusion

Within the next 20–30 years, agglomeration and diffusion will still be the important manifestation of the spatial organization of Chinese industries. Thus, to grasp the process, characteristics and mechanism of industrial agglomeration and diffusion and analyze the spatial effects of industrial agglomeration and diffusion will be the key to correctly identify the formation and evolution of industrial spatial structure. The research contents are: Firstly, to study the factors affecting industrial agglomeration and diffusion and their working mechanism, for example the influence of new factors such as institution, informatization, improvement in science and technology, industrial transformation, consumption mode, environmental ethics, *etc*. on industrial agglomeration and diffusion, and the relationship between the process of industrial agglomeration and diffusion and the formation and evolution of industrial spatial structure. Secondly, to explore the types of industrial agglomeration areas and the mechanism of industrial spatial integration, including the internal mechanisms, laws and trends of the formation and development of the key industrial agglomeration areas in China. Thirdly, to study the characteristics and typical spatial forms of industrial agglomeration and diffusion in regions at different spatial scales, and the intensity of industrial agglomeration and diffusion, including the regional differences of industrial agglomeration and diffusion, and the similarities, differences, and laws of the influences of industrial agglomeration and diffusion in different regions on industrial spatial restructure. Fourthly, to probe into the spatial relationship between industrial agglomeration and diffusion and regional transfer of production elements, including the effects of industrial agglomeration and diffusion on regional innovation, regional division of labor, and regional growth.

A1.2.3 Overall Goals and Stage Objectives to 2050

(A) Overall Goals

The key issues around this filed include industrialization, industrial distribution and industrial agglomeration. The overall goals are to reveal the trends, generative mechanism and evolution characteristics of spatial organization of industries under the intertwined influence of resource, environment and industrial development, to discover the new factors affecting industrial location decision and the trend of industrial distribution, to simulate the evolution of spatial structure of industries at different spatial scales and different development stages, to predict the trend of industrial distribution in China during the period of 2020–2050, and to provide scientific basis for the strategy-designing and policy-making of industrial development by the central and local governments.

(B) Stage Objectives

The major objectives to 2020 are to optimize spatial organization of industrial production on the condition that a rational growth of industrial economy is guaranteed. The sub-objectives would be to identify the existing and possible factors affecting and restricting the industrial activities in China through studying the main affecting elements of the spatial organization of industries and their working mechanism, and to provide security pre-warning for the threshold constrained by some typical elements; to put forward reasonable proposals for spatial organization of industries.

The major objective from 2020 to 2030 is to enhance the global competitiveness of Chinese industries through optimizing and adjusting the spatial organization of industries. There are some tasks to do, for example, the modes and mechanism of industrial production organization under the background of globalization, the main working mechanism of industrial agglomeration and diffusion, the relationship between industrial organization and regional development, *etc.*; the establishment of the theoretical system of industrial production organization with Chinese characteristics.

The major objective from 2030 to 2050 is to construct methodology, theory of industrial industry geography, *i.e.* researches on important mechanisms, theories and methodologies (App. Table 1.1).

App.Table 1.1 Roadmap of researches on China's spatial organization of industries

Projects	2020	2030	2050
Discrimination and analysis of main affecting factors	Discrimination of the main restricting factors and the restriction level, calculation of their safety threshold	Discrimination of gradual changing and reformative factors	Discrimination of optimizing factors
Working mechanism and mode of factors	The affecting mechanism of factors on spatial organization and connection	Mechanism of industrial agglomeration and diffusion	Theoretical system building of industrial agglomeration and diffusion
Interactive relationship between spatial organization of industries and regional development	Spatial representation and measurement of industrial parks and industrial bases	The interaction between spatial organization of industries and regional development	Coordination and regulation of regional spatial organization

A1.2.4 Implementation Plans and Technological Roadmap

Spatial organization of industries is the important research filed of economic geography. The research trend of the former is directed by the overall research contents and research methods, and at the same time has impact on the development of economic geography. Besides, industry, as the important support to regional economy, is closely connected with regional factors and the process of regional development. Therefore, it is of great significance to design a roadmap for the researches on spatial organization of industries.

The researches on spatial organization of industries should include the following four aspects: to study elements and working mechanism, and to explore spatial effects and institutions. These are four closely related and equally important aspects to spatial organization of industries. Therefore, the systematic research should be carried out from the four perspectives:

Firstly, to identify the elements leading to the changes of spatial distribution and linkage of industrial production and sales. Specifically, to analyze the progress of production technology and the changes in factor inputs from the angle of industrial production organization, but also probe into the non-production and non-technology factors, such as political, environmental, and social factors, *etc.* and their impacts on spatial organization in view of the increasing interdependence of global economy. Among these elements, some are restricting, such as resource and environment; some are reformative, *e.g.* the advanced technology may totally change the current factor input structure; and some are of gradual influence, such as political, social and institutional elements. Our short-term goal is to analyze the restricting elements, and the long-term one is to predict and identify the gradual changing and reformative elements in the long term.

Secondly, to analyze the affecting mechanism of factors' change on the production spatial organization, focusing on the way in which elements such as transportation and communication technology, tele-information technology, resource and environment, labor (skill, size and regulation), social culture, policy and institution affect the process of industrial production activities and spatial connection. This study will move beyond the traditional research methods in which only the black box connection between the characteristics of driving factors and their final spatial representations is described, and focus on process analysis instead. It will explore the influence of factor-driven on the element input, structure, spatial allocation and spatial connection of the industrial sector by means of cost-benefit analysis, production chain analysis, behavior-decision analysis and production network analysis.

Thirdly, to explore the formation mechanism and spatial effects of the spatial organization patterns of great significance to regional development. At present, these organization patterns mainly include industrial cluster, industrial agglomeration, comprehensive industrial base, *etc*., so this study will analyze the inner mechanisms of their formation and development, probe into their spatial representation, investigate the degree of agglomeration and dispersion, and discus their industrial and regional characteristics.

Fourthly, to discuss how to strengthen the interaction between industries and regions in order to provide better institutional environment for the optimization of spatial organization of industries. The study focuses would include the effects of adjustment and optimization of policies and institutions on industrial activities and spatial structure.

According to the importance and feasibility of the possible research questions in the four aspects, the research focuses of different stages are set as follows (App. Fig. 1.2).

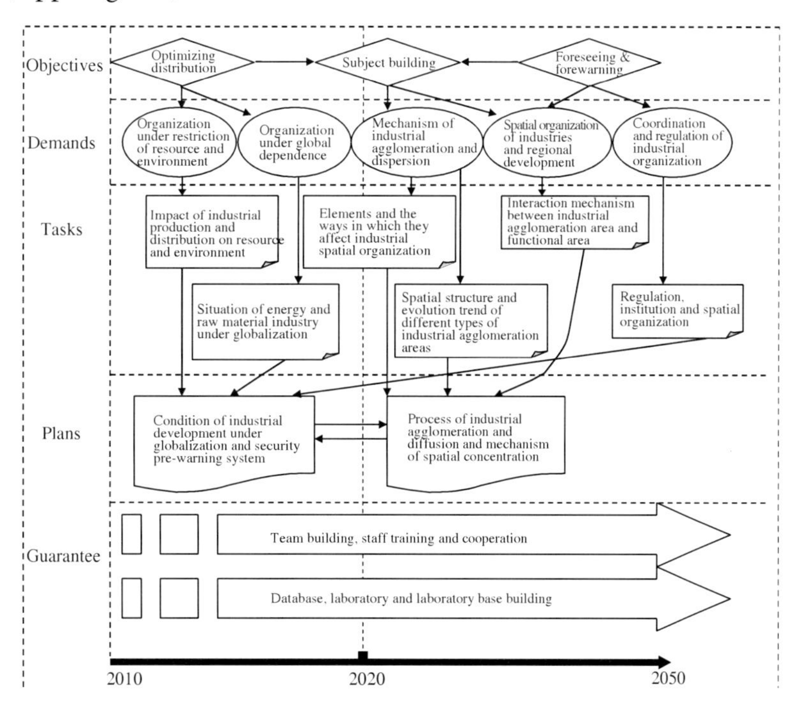

App. Fig. 1.2 Roadmap of researches on spatial organization of industries

(A) Research Focuses and Plan from Now to 2020

The present main research goal is to optimize spatial organization of industries. The main contents include reviewing critically the existing theories and approaches and forming the research framework, studying the factors which deeply restrict industrial production activities and the degree of their restriction, and foreseeing the trend and situation of the development and distribution of key industrial sectors such as raw-material and equipment industries.

Main research plan—the situation of industrial development under globalization and security pre-warning system.

To identify the major factors restricting spatial optimization of industries and on the basis of the quantitative analysis and comparison of the industrial development situation and competitiveness of different regions. In view of the increasing dependence of local development of industries on global market, this study would focus on the trends and safety coefficient of the industrial development and spatial organization in China, and on discriminating the major factors threatening Chinese industrial development and their dangerous threshold, and establishing security pre-warning system for industrial development.

Main contents:

- to analyze the industrial development trend and competitiveness;
- to study the spatial distribution of Chinese industrial production under the restrictions of resource and environment(water, soil, mineral resources, and ecologic environment);
- to explore the situation and mechanism of dependence of Chinese industries on global market;
- to investigate the development paths and spatial development patterns of energy, raw-material, and equipment industries under the background of globalization;

(B) Research Focuses and Plan from 2020 to 2050

The long-term research goal is to construct the systematical theory, methodology of industry geography, *i.e.* to explore working mechanisms, theories and methodologies concerned.

Main research plan—study on the process of industrial agglomeration and diffusion and the mechanisms of spatial concentration

The study will carry out comprehensive and integrative researches on the process of industrial agglomeration and diffusion and on the laws and mechanisms of the formation and evolution of industrial spatial structure from spatial, temporal and structural perspectives. From the structural perspective, the research focus is on the formation laws and evolution mechanisms of industrial structure. From the temporal perspective, the research will pay attention to the dynamic process of industrial spatial structure. From the spatial perspective, the research will focus on the characteristics, formation and transformation laws of industrial spatial structures at different spatial scales on the one hand, and the differences in the industrial spatial structures of different regions at similar spatial scales, and the nature and intensity of industrial spatial interaction on the other hand.

Main contents:

- to measure the level of industrial agglomeration and diffusion and its affecting factors;
- to delimitate different types of industrial agglomeration areas and study their spatial structure;
- to explore the evolution laws and mechanisms of industrial agglomeration areas;
- to develop the policies of promoting different industrial agglomeration areas;
- to investigate interactive mechanisms of industrial agglomeration areas and functional zones.

A1.3 China's Agricultural and Rural Development Roadmap

A1.3.1 Status Quo and Research Progress of China's Agricultural and Rural Development

Rural region is essentially a geographical system, meaning all areas other than urban regions[100,101]. In 2006, China's urban built-up area amounted to 33, 700 km^2, and the township area was less than 23, 000 km^2, with the rest, rural region accounting for 99.38% of China's total land area. China's rural population in 2006 accounted for 56.1% of the total population. Therefore, China is not only a nation with a huge population of 1.3 billion, but also a large agricultural country.

China's Agricultural and Rural Development (CARD) had experienced four major stages from 1949 to the early 1980s[102]: 1) the stage of rapid agricultural recovery (1949–1955); 2) the stage of agricultural stagnation and recession (1956–1962); 3) the stage of slow agricultural resurgence (1963–1980); and 4) the stage of rapid agricultural development (the early 1980s). With the complete establishment of basic rural management institutions, China's rural development has entered an important stage of exploring marketization reform with emphases on the reform of agricultural product circulation system, the fostering of agricultural products market, the strategic adjustment of rural industrial structure and the promotion of the development of non-agricultural enterprises, and so on. These measures not only resulted in the emergence of the township and village enterprises (TVEs) in the mid 1980s, but also promoted the transformation of rural economy from the traditional unitary structure of agricultural industry to the integrated multi-sector structure.

The rural development of China has entered a period of overall transformation towards the socialist market economy since 1992. With the development of agriculture and rural economy stepping into a new stage, complex issues and contradictions related to rural development were exposed and have been increasingly concerned. In order to solve these problems and promote the coordinated development of urban and rural areas, the central government of China mapped out an important strategy of "the new socialist countryside construction" in 2006, by developing production, improving living standards, fostering more civil behavior, improving the overall cleanliness of villages, and exercising democratic management. Recently, positive measures have been adopted to enhance the security level of the quality and the comprehensive international competitiveness of agricultural products, to increase the rural productivity and peasants' income, and to adjust the industrial structure so as to improve the agricultural management level and economic benefits through carrying out industrialization management.

Before the transformation towards market economic system, researches on CARD mainly focused on regional planning on the status quo of agricultural development[103], comprehensive regional planning of agricultural development[104] and rural settlements geography[105-107], *etc*., and *China's Agricultural Geography*[108], *Rural Settlements Geography*[109], and *Rural Geography*[110] were the outstanding works. Thereafter, the research focuses have been shifted to the sustainability[111] and challenges[112,113] of agriculture, the sustainable development of agriculture and rural economy[114-116], dual rural-urban relationship[117], village labour migration[118], rural reform and urban-rural coordinated and balanced development[119], *etc*.

With the implementation of the strategy of "the new socialist countryside construction," related researches have been widely concerned in the academic circle[120,121]. On the basis of the research concerning new countryside construction in eastern coastal China, Prof. Yansui Liu put forward an innovative view of "three conformities and one upgrade" to promote rural development, *i.e*., spatial-territorial conformity aiming at improving rural land-use efficiency by carrying out rural construction land consolidation; primary administrative unit conformity aiming at pushing forward urbanization and improving modern community service and management level by strengthening the construction of central village and rural new community; rural industry conformity aiming at developing internationalized modern agriculture and joint-stock enterprises and developing circular economy. All these measures are taken to upgrade rural productivity and competitiveness[122]. Currently, regional differences, rural poverty, rural land-use issues and international environment are the four major influencing factors of the building of a new countryside in China[123]. Accordingly, some researches, such as the implication of rural land-use change on the building of a new countryside[124,125], rural development types and their regional differentiation driven by industrialization and urbanization[126] and so on, are carried out in depth.

A1.3.2 Trends of CARD and Scientific and Technological Demands

(A) Factors Affecting CARD Will Exist for A Long Time and Some Contradictions Will Be Intensified.

To some extent, economic globalization means the re-allocation of resources across the world. However, in this process, the optimization and allocation of the three major factors influencing CARD, *i.e.* land, labor and capital, are constrained. These factors will potentially impede the further development of China's agricultural modernization[127]. Rural land system will be restricted by: 1) ambiguous property rights of rural land led to the lack of effective protection of farmland property right; 2) instability of the contract right affects peasants' investment anticipation and leads to short- term opportunism behaviors of rural land-use; 3) unsmooth land transfer results in low efficiency of farmland allocation. Low comparative benefits from agricultural production have resulted in the transformation of massive agricultural production factors to non-agricultural industries, which shrank peasants' enthusiasm on agricultural sector. Relatively low quality of peasants and tremendous pressures on resources and employment will become the major obstacles for future CARD which will also be influenced by rural depopulation, counter-urbanization[128] and rural land system reform.

(B) CARD Is Increasingly Oriented Towards Modernization, Internationalization and Marketization

The future directions of CARD are concluded as follows: 1) to focus on advancing production and enhancing economic benefits; 2) to pay more attention to clean agricultural production and food security and to protect rural ecological environment; 3) to participate in international trade and to quicken the internationalized development of agriculture based on comparative advantages; and 4) to develop rural economy with distinguishing features on the basis of regional differentiation. With the development of urbanization and industrialization, the relationships between urban and rural areas and between industries and agriculture will be totally changed, and China's agricultural and rural economy will enter a stage of transitional development with the characteristics of transforming from "being deprived" to "being supported".

The future trends of China's agricultural development mainly include: 1) Chinese agriculture will be exposed to the intensified competition of global market forces; 2) strategic adjustment of agricultural structure will be implemented; 3) agricultural development strategies focusing on regionalization, base-construction and modernization will be improved; and 4) institutional reform aiming at marketization,scale management and peasant economy will be inevitably carried out. The 30-year household contract responsibility system will be gradually turned to the coexistence of multiple operational modes of economic elements owing to the rise of agricultural joint-stock system, professional cooperative organizations and modern agricultural park.

(C) CARD Will Become the Important Basis for Establishing A New Pattern of Rural-urban Integration

Following the principles of "regional comparative advantages" and "effective market demand", the adjustment of the agricultural structure in China has become fundamental driving force of promoting China's agriculture toward regionalization, specialization and marketization. In this context, to 2030, the advantageous agricultural production area and new industrial belts will be formed; sustained and highly efficient modern agriculture will be dominant in Chinese agriculture.

There is a close relationship between future rural development pattern and rural-urban transition development. The characteristics of future rural-urban development in China will be represented in the aspects of spatial-territorial reorganization and industrial restructuring: 1) industrial restructuring will lead to the transformation of traditional rural industry into urban and modern garden industry in order to realize the industrial connections between rural and urban areas; 2) spatial integration of rural and urban areas will be realized with the rural population migrating to the cities and settling down there as well as the trend of counter-urbanization; 3) the dualistic rural-urban structure will be broken down, and the value of agricultural factors and rural area will be highlighted due to the orderly and effective flow of rural and urban factors abiding by the market law; and 4) the mechanism and platform of rural-urban equality and rural-urban integration will be established. The core of the connection and interactive between rural and urban economy is to enhance the comprehensive strength and competitiveness of rural development. The construction of "a new countryside" provides solutions to the various problems in the balanced and coordinated development of rural and urban areas and thus will become the mainstream of the harmonious economic and social development of China in 30–40 years.

(D) CARD Going Towards to Integrated and Systematic Researches

There is an urgent need for innovative researches on CARD to provide strong support to realize coordinated and balanced rural-urban development, the balance between agricultural economy and social issues, and appropriately deal with the relationship between rural development and environmental protection. Therefore, to meet the needs of national strategies, many complex and integrated scientific propositions are put forward for the innovative researches on CARD, which include: 1) the new patterns of regional agricultural types; 2) rural-urban relationship and spatial structure; 3) driving mechanism and competitiveness of rural development; 4) the effects of rural factors transformation; 5) differentiation and restructuring of rural territorial types; 6) multi-functionality of rural system during transition period; and 7) the mechanism and comprehensive improvement model of rural settlement hollowization.

A1.3.3 The Overall Goals and Stage Objectives to 2050

Taking into consideration of the above scientific and technological demands, the deep-seated contradictions and development issues during the transition period of Chinese rural development need to be studied in different stages and different ways according to their different degrees of urgency so as to provide scientific-technological and policy support to construct Chinese new countryside and realize the coordinated rural-urban development. To 2050, the main objectives of innovative researches on CARD are: 1) the differentiation and restructuring of rural territorial types; 2) the diversity of rural system functions and the formation mechanisms of rural settlement hollowization; 3) the new patterns of regional agricultural types and the effects of rural factors transformation; 4) rural-urban relationship and spatial structure pattern; and 5) driving mechanism and competitiveness of rural development, *etc*.

(A) The Research Subjects and Objectives (from 2010 to 2020)

a. The Mechanism of Rrural Settlement Hollowization and Its Improvement

To deal with the phenomenon of rural settlement hollowization in the process of rural resources reallocation and non-agricultural transformation of rural population led to by the long existing dualistic structure of rural-urban resources allocation, and to explore regional differentiation, lifecycle and socio-economic dynamic mechanism of the rural settlement hollowization, and to search for corresponding countermeasures.

b. Differentiation and Restructuring of Rural Territorial Types

To discover the main controlling factors of rural spatial-territorial differentiation and their interaction mechanism, to identify the rural spatial-territorial types and their spatial structure, and the main functions of different rural spatial-territorial types, and to explore the coupling process and mechanism of the main function-oriented rural-urban spatial-territorial system.

c. Diversity of Rural System Functions during the Transition Period

To study the rural spatial-territorial differentiation and functional expansion as well as how to promote rural development on the basis of the multiple function of rural system and local abundant natural resources and unique rural culture, and finally achieve the smooth transition of rural industry.

(B) The Research Subjects and Objectives (from 2020 to 2030)

a. The New Patterns of Regional Agricultural Types

To study the new patterns and development process of agricultural advantageous areas, industrial belts, and agricultural security production in a new context of economic globalization and emphasizing the main function, to simulate and analyze the evolutionary process of transporting north grain to the south, and its comprehensive effects on ecological environment, water resource and rural development in north China, and to put forward optimizing countermeasures on the regional allocation of agricultural resources so as to provide scientific basis for the planning- designing and policy-making of sustainable CARD.

b. The Effects and Value of Rural Factors Transformation

To study the process, spatial agglomeration and effects of the non-agricultural transformation of rural factors, *e.g.*, land, capital and labor, under the condition of market economy, with the integrated development of rural and urban areas, so as to provide scientific basis for the integrated rural-urban socio-economic development and the optimization of the rural-urban spatial-territorial patterns.

(C) The Research Subjects and Objectives (from 2030 to 2050)

a. Rural-urban Relationship and Spatial-territorial Structure

In view of the trend of transformation from the dualistic structure to the rural-urban integrated development, we will carry out in-depth researches on the possible socio-economic issues in the process of rural-urban transition and the integrated development, in order to provide powerful scientific-technological support for the scientific planning and administration of promoting the rural-urban integrated and coordinated development.

b. Driving Mechanism and Competitiveness of Rural Development

Taking into consideration the weak market adaptation and self-development abilities of Chinese rural development owing to long-term benefit deprivation, we will carry out in-depth researches on the mechanisms and laws of rural transformation development under the pressure of rapid industrialization and urbanization process as well as on the dynamic mechanism of rural revitalization and rural-urban coordinated development, and to put forward diversified models and path for improving the comprehensive productivity and international competitiveness of Chinese agriculture.

A1.3.4 Implementation Plans and Technological Roadmap

(A) Major Scientific Issues and Research Contents

a. The Mechanisms and Governance Model of Rural Settlement Hollowization

Research contents include: spatial-temporal characteristics of the non-agricultural transformation process and employment transfer of Chinese rural population; the life cycle of rural settlement hollowization in different types of regions; the scientific improvement model for rural settlement hollowization; the scale and spatial structure of rural villages in different types of regions; and the innovative mechanism, operation models and optimizing paths of improvement on hollowing village, rural housing, and farmland reclamation in different types of regions.

b. Differentiation and Restructuring of Rural Territorial Types

The main controlling factors of rural spatial-territorial differentiation and their interaction mechanisms; the identification of the rural spatial-territorial types and their spatial structure; the main function of different types of rural territories; the coupling process and mechanism of the main function-orientated rural-urban territorial system; and the spatial-territorial pattern of "three conformities and one upgrade" for rural restructuring.

c. Multi-functionality of Rural System during the Transition Period

The rural spatial-territorial differentiation and its functional expansion in aspects of rural ecology, rural culture, space, *etc*.; rural development on the basis of the multi-functionality of rural system and of local abundant natural resources and unique rural culture.

d. The New Patterns of Regional Agricultural Types

The new patterns and development process of agricultural advantageous areas and industrial belts; simulation and analysis of the evolution process of transporting north grain to the south, and its comprehensive effects on ecological environment, water resource and rural development in north China; modeling China's new modern agricultural pattern with the characteristics of "vast territory, huge amount of food and large-scale agriculture".

e. The Effects and Value of Rural Factors Transformation

The value-increased mechanism of the non-agricultural transformation process of rural elements in the context of urban and rural equalization; systematic evaluation on the process, intensity and effects of the influence of non-agricultural transformation of rural elements on rural sustainable development at the village and household levels; the rural spatial restructuring under the comprehensive influence of the non-agricultural transformation of rural elements and favorable policies.

f. Rural-urban Relationship and Spatial Restructuring

How to well understand the equal status of rural and urban areas and the dynamic rural-urban relationship; how to scientifically judge the interaction mechanism between rural and urban systems and their coordination threshold; how to implement rural-urban spatial restructuring based on the coupling relationship among resource, environment and industrial development; and how to promote the scientific planning and management innovation of rural-urban integrated development.

g. Driving Mechanism and Competitiveness of Rural Development

How to follow the law of rural transformation development under the pressure of rapid industrialization and urbanization process and to promote the self-organizing mechanisms of rural revitalization and rural-urban coordinated development, including self-development mechanism, interactive mechanism for nurturing and fair play and harmonious mechanism, *etc*.; how to adapt to the economic globalization process and to put forward comprehensive paths for improving the economic strength and international competitiveness of Chinese agriculture and rural areas based on the comparative advantages and market rules.

(B) Expected Projects

a. Project Title

- Regional differentiation;
- Dynamic mechanism and improvement model of rural settlement hollowization in China.

b. Research Objectives

Employing geographical theories and methods, and other disciplines such as resources science, economics and information science, this project is to reveal the law and evolutionary mechanisms of rural territorial system during the rural-urban transition process, to comprehensively analyze the dynamic process and life cycle of rural settlement hollowization, and to discover the main driving forces and effects of rural settlement hollowization in different types of regions, to explore the regional models for improving rural settlement hollowization and paths for the sustainable development of new countryside construction, and to provide scientific and policy basis for national strategies on the coordinated rural-urban development and new countryside construction.

c. Major Research Contents

(1) To study the dynamic mechanism, development stages and territorial types of rural settlement hollowization.

(2) To analyze the evolutionary model and types of hollowing village in the process of urbanization, the spatial-temporal characteristics of the expansion of hollowing village at specific economic development level and in specific geographical environment, and its impacts on the social and spatial relationship in villages.

(3) To study the development stages of hollowing village, its diagnosing indicators, and standards for the classification of the hollowing villages in typical regions, and the development process and differentiation laws of hollowing village in specific region.

(4) To study the socio-economic dynamic mechanism of the formation of abandoned and idle lands in hollowing village, and the effects of the rural settlement hollowization on resources, environment and social economy.

(5) To study the standards and criteria for selecting central villages and designing their distribution pattern.

(6) To study the regional model of rural settlement hollowization improvement, and the planning technology for rural spatial restructuring; and to innovate the collaboration mechanism between scientific support and government policies.

(C) CARD Roadmap (App.Fig.1.3)

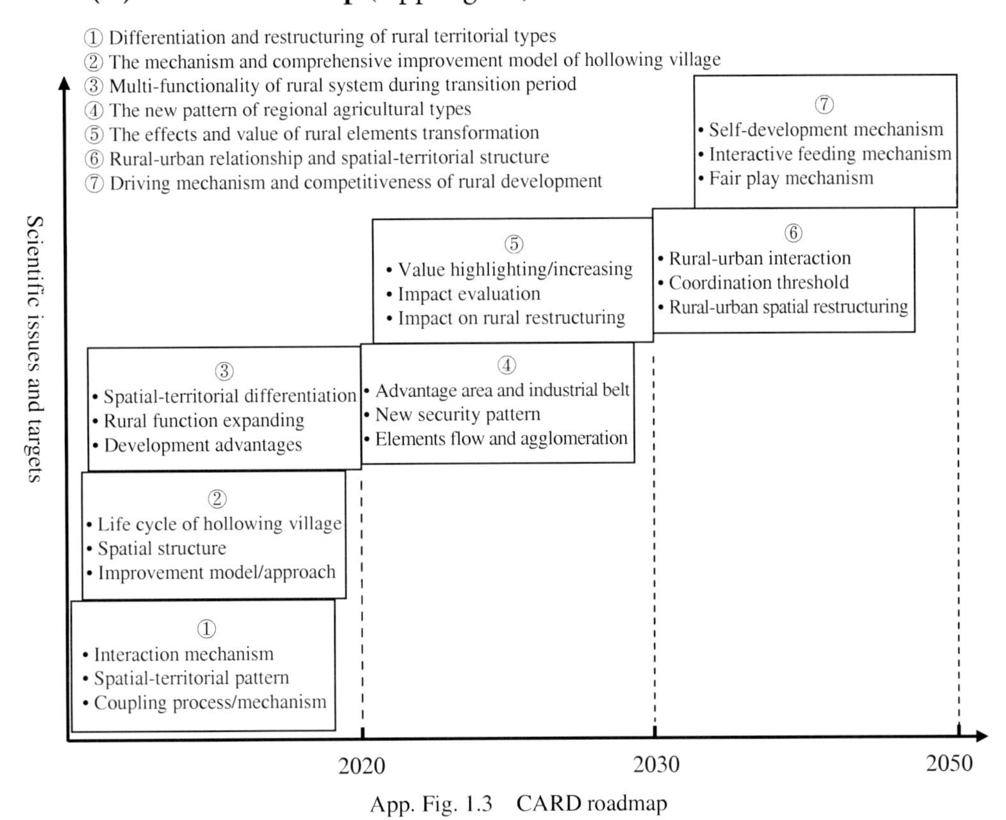

App. Fig. 1.3 CARD roadmap

A1.4 Prospects and Spatial Pattern of China's Urbanization

The urbanization process in China is exerting profound influence on the world as well as on China in the era of economic globalization. Toffler, the world's leading futurologist, believes that the development of high-tech industry and China's urbanization are the two main factors driving global economy and social development in the 21st century. Healthy urbanization development is both a strategical historical mission in the construction of modernization and a significant guarantee for building a harmonious society[129]. Thus, we are supposed to unswervingly persist in taking urbanization as one of our national development strategies. In the next several decades, urbanization will experience a rapid development, and there will be a large number of rural labor forces immigrating to urban areas. The shift from low productivity to high productivity will be part of the advancement of human society. China's urbanization will not only shape the future of China, but also have crucial effects on the world urbanization process[130].

A1.4.1 Overall Evaluation of China's Urbanization

China's urbanization has experienced a 57-year tortuous development process. After detailed analysis, we can see that China's urbanization approximately conforms to the "S" curve's rule since 1949, as is shown in other countries around the world. China spent 47 years in completing the taking-off stage of urbanization, and entered the stage of rapid development after 1996 (App. Fig. 1.4).

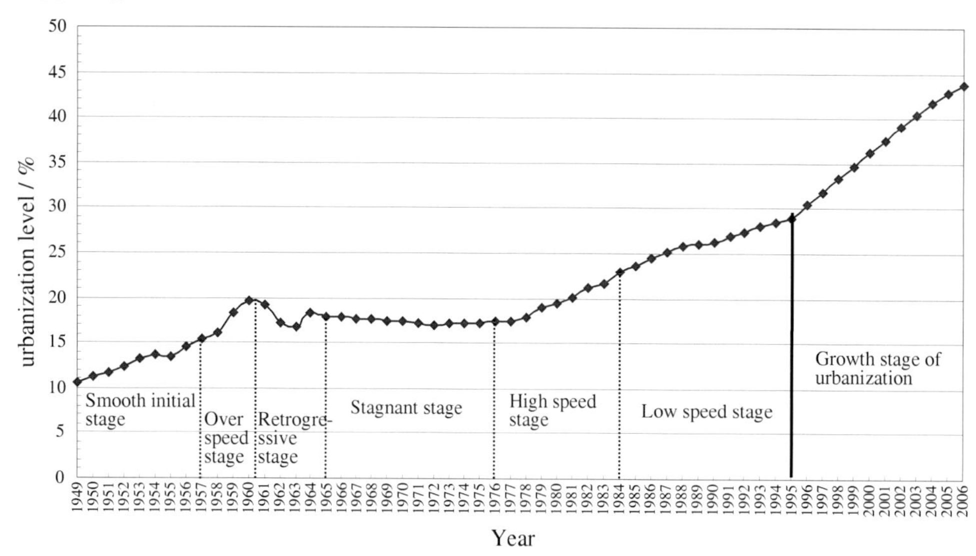

App. Fig. 1.4 Growth and development stages of national urbanization level since 1949

Especially in the 30 years of reform and opening up, great achievements have been made in the urbanization development and urban construction. During this splendid period of time, China's urban population has risen from 173 million in 1978 to 594 million in 2007, with the average annual growth rate of 4.4 percent. At the same time, the urbanization level has improved from 17.92% to 44.99%, with an average annual increase rate of 0.93%. Furthermore, the growing speed of urban population is 2.14% faster than the world average. However, the present urbanization level of China is still 8.40% lower than the world average. The number of cities in China has increased from 190 to 661, with an average annual increase of 16. Meanwhile, the size and number of towns are expanding, the latter one increasing from more than 2000 to about 18.9 thousand, with an average annual increase of 640[131].

With the enhancement of urbanization level, the urban infrastructure construction has made momentous progress, urban public service facilities have been constantly improved, and the living standard and civilization level of urban residents have also been greatly enhanced. During the past several decades, the unceasing adjustments of the path of China's urbanization effectively promoted its healthy development. The urban system with reasonable industry division of labor, orderly spatial competition, and an efficient market was preliminarily established. The rationalization of urban system made urban agglomerations become the basic geographical units to participate in the global competition and division of labor, the experimental areas of national comprehensive reforms, and the determining factors of China's economic development[132]. Meanwhile, the urbanization theory was improved continuously, and more and more related research institutions and talents emerged in this field.

Along with the great achievements, a series of urgent and practical problems inevitably appear in the process of urbanization. To be specific, there are problems mainly in four aspects. First of all, the speed of urbanization is too fast and falsely high, and the phenomenon of rash advance of urbanization development occurs in many places[133]. Secondly, the expansion of urban space has been out of control to some extent. Thirdly, the economic development and employment capacity cannot meet the demands of rapid urbanization development. Finally, the resource and environmental deprivation and conservation in the process of urbanization bring more and more pressure on resources and ecological environment, and further widen the gap between the rich and the poor[134]. Whether these problems can be solved in a new round of development or not will directly affect the prospect of China's urbanization in the next few decades.

So far as the researches on urbanization are concerned, the previous ones mainly focused on the exploration of the path of urbanization with Chinese characteristics and the enhancement of the speed and level of urbanization, *i.e.* from the perspective of "quantity" during the past three decades; while the "quality" of urbanization was not been paid due attention to, including the efficiency and welfare of urbanization, people's living standard, the level of civilization, the protection of resources and ecological environment, the improvement infrastructure, the urban employment and the development of service industries, *etc*.

A1.4.2 Overall Judgment on Trend and Pattern of Urbanization and the Scientific and Technological Demand Analysis

(A) Basic Judgment on the Future Development Trend of Urbanization

(1) China's population to 2050. According to the multi-state population projection model (PDE model), the national population will hit 1,425 million by 2020, reach the peak of 1,442 million in 2030, and by 2050 it will reduce to 1,383 million. In the next 50 years, there will not be obvious changes in the general trend of the population distribution of China. After 2015, China will enter a period of rapid development of the aging of population. At that time, the elderly population will account for 20% of the total national population, while the proportion of 0–14 year-old children will drop to 15% by 2050.

(2) China's floating population to 2050. According to the historical trend and the future population migration model, in the next 50 years, the proportion of long-distance floating population will increase to 35% by 2020. The proportion of short-distance floating population will decline to about 30%, and

the proportion of medium-distance floating population will rise to about 35%. The eastern region will continue to be an area with net inflow of population, the central region will have a balanced population inflow and outflow and the western region will still have a net population outflow.

(3) China's urbanization level to 2050. The urban population of China will reach 775 million in 2020, 888 million in 2030, and 940 million in 2050. And the corresponding urbanization level will rise to 54.45%, 61.63%, and 67.98% respectively. The annual increase rate of urbanization level will be 0.89% from 2005 to 2020, 0.68% from 2020 to 2030, and 0.32% from 2030 to 2050. From 2030, China's urbanization will enter the stage of slow growth and mature development.

(4) To 2050, the situation of water and land resources guarantee will be quite severe in the process of urbanization. According to the analysis of the variation mechanism and laws of the guarantee degree of urban water use and construction land use, 1.7 billion m^3 urban water and 1,004 km^2 urban lands were used for every 1% increase of the urbanization level in the past 25 years. And in the coming 15 years, the figures will be 32 billion m^3 and 3,459 km^2 respectively. This indicates a harsh situation of water and land resources guarantee in the urbanization process in a few decades. There is a trend that the degree of urban water and land use guarantee gradually decline geographically from western region to eastern region[135].

(5) To 2050, the situation of eco-environmental security will also be severe in the process of urbanization. To 2050, every 1% increase in the urbanization ratio will entail the increase of 105 million hm^2 total ecological footprint, 0.2 million hm^2 construction land ecological footprint and 11 million hm^2 per capita ecological footprint, while the intensity of the ecological footprint will drop by 6 hm^2/yuan, the ecological overload will increase by 5.68%, and the eco-environmental quality index will decline by 0.0064. In the next few decades, following the existing development mode of urbanization, the ecological overload will become more serious and the eco-environment will continue to deteriorate.

(6) The path of China's urbanization to 2050. In view of the requirement of the resource and environmental carrying capacity, we should make efforts to establish the healthy urbanization pattern with efficient resource, friendly environment, effective economy, and harmonious society, and to find a compacted, highly efficient, saving and differentiated path of urbanization with Chinese characteristics according to the principles of "step-by-step, coordinated urban and rural development, intensiveness and high efficiency, adaptation to local conditions, and multiple promotion."

(B)Future Spatial Patterns of the Reorganization of National Urban System

(1) To 2050, a general urban development pattern in China with a coordinated urban and rural development will be formed which is composed of 3 mega urban agglomerations, 20 urban agglomerations, 5 ultra-mega-cities, 10

mega-cities, 50 major cities, 150 large cities, 200 medium-sized cities, 250 small cities and towns connected by 5 national urban development belts.

(2) To 2005, a national urban agglomeration system consisting of 23 urban agglomerations such as the Yangtze River Delta mega urban agglomeration, the Pearl River Delta mega urban agglomeration, the greater Beijing area mega urban agglomeration, Liaodong Peninsula urban agglomeration, Shandong Peninsula urban agglomeration and so on will be formed. And a group of innovative, ecology-based, low-carbon, digital mega urban agglomerations and large urban agglomerations will be constructed. Urban agglomeration will become a new geographical unit participating in the global competition and international division of labor, the most dynamic core strategic growth pole with the greatest potential which will profoundly affect the international competitiveness of China, and the determining factor of China's general strategic situation of modernization and urbanization.

(3) To 2050, a national urban system composed of 5 super-mega-cities (Beijing, Shanghai, Tian'jin, Guangzhou, Nanjing) with over 10 million population in each city, 10 mega-cities (Wuhan, Chongqing, Chengdu, Shenyang, Zhengzhou, Xi'an, Shenzhen, Jinan, Qingdao, Harbin) with over five million population for each one of them, 50 major cities with the population of each city over 1 million, 150 large cities with 0.5 million population in each city, 200 medium-sized cities, and 250 small cities will be formed in China.

(4) To 2050, a main framework system of Chinese urban development composed of "three vertical and two horizontal" belts will be established. The "three vertical" belts include the eastern vertical urban development belt, central vertical urban development belt, and western vertical urban development belt, while the "two horizontal" belts are the Yangtze River horizontal urban development belts and the Eurasian Continental Bridge horizontal urban development belts[136].

(C) Scientific and Technological Demand of National Urbanization Development and Pattern Formation

In view of the general future trend and spatial pattern of China's urbanization, we should study the scientific and technologic demands according to the following facts which form the main clue for the analysis: unswervingly persist in taking urbanization as one of our national strategies; the urbanization process is composed of several irreversible stages; China has irresistibly entered the fast growth stage of urbanization; it is difficult to solve the living problems of the rural migrant workers; we have to work out the ultimate destination of China's urbanization; we should spare no efforts to find out the path and modes of China's urbanization conforming to China's actual conditions; no prediction can be made on the changes in the reconstruction patterns of China's urban spatial structure. To be specific, the following contents are included:

(1) The scientific and technological demands of speed control and quality upgrade in different urbanization stages under the background of globalization, measurement of the threshold of urbanization in different types of areas,

analysis of the affecting factors of urbanization and the relationship between these factors and the carrying capacity of resources and the environment.

(2) Scientific and technological demands of the coordinated development of urbanization, economy, and economic restructuring under the background of modernization.

(3) Scientific and technological demands of solving the problems induced by further urbanization such as resource and environment deprivation and protection, employment guarantee, energy guarantee, and social security problems, the measurement of the guarantee degree of these problems, and the mechanism analysis.

(4) Scientific and technological demands of answering the following questions: How will the spatial structure of urban system be reorganized in the future? What kinds of impact will the reorganization have on the overall productivity layout of China? How will the reorganization affect the structure of natural environment and the reorganization of land space?

(5) Scientific and technological demands of selecting the paths and modes of urbanization corresponding to the different urbanization stages in the future. How will the pattern of Chinese urban spatial structure change under the influence of the path and mode of urbanization in the future? How will the pattern of China's urban spatial structure be adjusted to response to the resource and environmental guarantee in the future?

A1.4.3 Overall Goals and Stage Objectives to 2050

(A) Overall Goals and Stage Objectives in Urbanization process

The overall objective of the urbanization development and its spatial patterns in China is to deal with the challenges brought by the era of globalization and information. To be specific, to implement the Scientific Outlook on Development, increase the level of urbanization to about 68% based on the restriction of the carrying capacity of resource and environment, continuously improve the quality of urbanization development so as to ensure the healthy and steady transition into the mature stage of urbanization in China, make efforts to form a brand new pattern of healthy urbanization development with efficient resource utilization, friendly environment, highly effective economy, and harmonious society, construct ecology-based, low-carbon, innovative, internationalized, informatized and modernized cities, and ultimately establish the main framework system of urban development and the structure system of cities and urban agglomerations.

(1) To 2020, the level of urbanization will have been increased to 54.45% with an annual increase rate of 0.89%; the medium-term growth stage of the urbanization process will be smoothly accomplished; according to the overall objectives, the urban structure system of China which consists of 2 ultra mega-cities (*i.e.*, Beijing, Shanghai), 3 mega-cities (*i.e.*, Tianjin, Guangzhou, and Nanjing), 20 major-cities, 50 large cities, 200 medium-sized cities and 350 small cities will have been established; 2 huge urban belts, *i.e.* the east vertical

urban belt and the Yangtze River horizontal urban belt, and 8 large-scale urban agglomerations will have been fostered and formed.

(2) To 2030, the urbanization level will have been increased to 61.63% with an annual increase rate of 0.68%; the urbanization of China will have stepped into the later mature stage; according to the above principles of urban construction, the urban structure system of China composed of 3 ultra mega-cities (*i.e.*, Beijing, Shanghai and Tianjin), 6 mega-cities (*i.e.*, Guangzhou, Nanjing, Wuhan, Chongqing, Shenyang and Zhengzhou), 30 major-cities, 100 large cities, 150 medium-sized cities and 300 small cities will have been established; 3 huge city-belts, *i.e.* the east vertical urban belt, the Yangtze River horizontal urban belt, and the central vertical urban belt, and 15 large-scale urban agglomerations will have been fostered and formed.

(3) To 2050, the urbanization level will have reached as high as 67. 98% with an annual increase rate of 0.32%; the urbanization of China will be fully developed in the mature stage; the national urban structure system of 5 ultra mega-cities, 10 mega-cities, 50 major-cities, 150 large cities, 150 medium-sized cities and 250 small cities will have finally been established; the main national urban development framework and urban agglomeration structure system composed of 5 huge urban belts and 23 large-scale urban agglomerations will have been formed.

(B) Overall Research Goals and Stage Objectives

In view of the situation of blind development of urbanization and the new problems in coordinating development of urban and rural areas at the present stage, the research points are to analyze the major territorial types and their basic features, key influencing factors and the evolution trend of China's urbanization on the basis of the analysis of the process, dynamic mechanism and future prospects of the urbanization; to compare the influences of the spatial configuration of different types of urbanization on resources and environment; to reveal the conditions forming various urbanization modes and the mechanism of coordinated urban and rural development through the case study of typical regions; to analyze the spatial pattern of the urban system re-construction and the spatial distribution pattern of urban agglomeration structure system in the context of globalization; to design optimization schemes for the spatial structure of urban system on the basis of homeland security; to develop evaluation, pre-warning and experimental computing systems for the sustainable urban development; to put forward the policy frameworks for urbanization regulation and coordinated urban and rural development, basic public service and social security system, and work out major paths to establish these systems.

(1) During the period of 2010–2020, the research focus will be the mechanism of urbanization. To study the dynamics and key factors influencing the urbanization process of China, to analyze the main territorial types and basic features of Chinese urbanization, and to predict the prospect of the urbanization and the threshold of the carrying capacity of resources and

environment, to compare the effects of the spatial configuration of different types of urbanization on resources and environment, to reveal the forming mechanism and scientific foundation of urban system and urban agglomeration structure system, and to select appropriate urbanization paths and modes in accordance with the actual resource and environmental conditions in China.

(2) During the period of 2021–2030, the research will focus on the process of urbanization. To dynamically simulate the overall evolution situation and trend of the urbanization process and urban spatial patterns, to analyze the degree of natural resources and environment guarantee in the process of urbanization, to reveal the process and paths of the material transfer between urban and rural areas as well as the optimal evolution pattern of urban and rural structure, to study the forming and growing process and the differentiation law of the urban agglomerations in China, and to explore the technical process and paths of the ecological, low-carbon and innovative urban construction.

(3) During the period of 2031–2050, the research will concern the spatial pattern of urbanization. To carry out reconstruction of the spatial pattern of urban system and urban agglomeration system in the context of globalization, to design optimization schemes for the spatial structure of China's urban system on the basis of the homeland security. Other significant research objects include: global urban areas, mega-cities and large urban agglomerations (urban belts), construction of comprehensive urban GIS and dynamic evolution simulation platform, policy system of the regulation on the urbanization of various types of regions and on the coordinated development of urban and rural areas, experimental computing system of the industrial clusters and spatial reorganization of urban agglomerations, construction of international metropolises, national and regional hub cities, development of the evaluation and early-warning systems for the sustainable urban development, technical specification and policy framework of healthy urban construction, basic public service and social security system under the new spatial pattern of urbanization.

A1.4.4 Implementation Plans and Technological Roadmap

(A) Key Scientific Issues

The research mainly focuses on the following key questions. First, to describe the urbanization in different scenarios, and carry out comprehensive and integrative study on the influencing factors of urbanization from the perspective of the coupling of pattern and process. Secondly, to predict the trend of key factors driving the ecological environment and those of urbanization by means of the integrated coupling model, and to measure the capacity of urban development in accordance with the dynamic change of ecological environment of typical areas and the effects of urbanization on the environment. Thirdly, to explore the regional pattern of rational urbanization with consideration on the coordinated urban and rural development according to the uniqueness of Chinese urban development. Fourthly, to make the urban dynamic evolutionary model better conform to the reality of Chinese urbanization development through introducing controlling factors and restrictions. Meanwhile a simulation platform for the evolution of comprehensive urbanization system and a policy support system platform of the sustainable urban development will be developed by employing GIS technology, massive spatial data management and computation platform.

(B) Major Research Contents and the Roadmap

According to the research objectives, the researches will be carried out following the main research line of the mechanism–process–pattern of urbanization in a chronological order with emphases on the following six major topics, and corresponding project proposals will be raised for each topic. App. Fig. 1.5 depicts the general view of main researches in successive phases.

(1) Research on the dynamic mechanism of urbanization, new factors influencing the regional pattern of urbanization, driving factors, dynamic monitoring/tracing/early-warning of future scenarios.

(2) Research on the impact of urbanization on resources and environment, the degree of natural resource and environmental guarantee and the threshold of the carrying capacity of resource and environment, technical realization of ecological, low-carbon and innovative urban construction.

(3) Research on the forming and growing mechanism and scientific foundation of urban (agglomeration) system, spatial differentiation law, and the optimal urbanization path and modes in accordance with the actual resource and environmental conditions in China.

(4) Research on the spatial pattern of urban (agglomeration) system in the context of globalization, optimization program for the urban spatial reorganization based on homeland security, global urban areas, mega cities and urban agglomerations (belts).

(5) Research on the material transfer, structure evolution, balanced development and policy system of/between urban and rural areas.

(6) Research on the evolution, pre-warning and policy supporting system and experimental computation platform of the sustainable urban development, the simulation of the urbanization process of key cities and regions, and the development of a comprehensive urban GIS platform.

(C) Technological Roadmap

To identify the different stages of China's urbanization development, the dynamic mechanisms and main features of different urbanization stages by means of reviewing literature and global comparison; to predict the development trend and spatial strategy selection of urbanization through multi-scheme models and scenario analysis; to classify different territorial types of urban development and urban-rural relationship combining the GIS technology, mathematical statistics and expert knowledge, and to analyze the key factors affecting the changes in territorial types. Furthermore, both qualitative analysis and quantitative models such as the Markov chain, CA model, *etc*. are employed to predict the evolution trend of the spatial pattern of China's urbanization; different types of spatial patterns of urbanization and the effects of urban-rural relationships under different types of urbanization on natural resources are compared through the comprehensive application of RS, GIS, GPS and mathematical statistics; dynamic simulation of urbanization evolution and comprehensive and integrated analysis are to be employed for the establishment of a policy support system for sustainable urban development, and the policy system and basic paths of China's urbanization and the balanced urban-rural development can be derived accordingly.

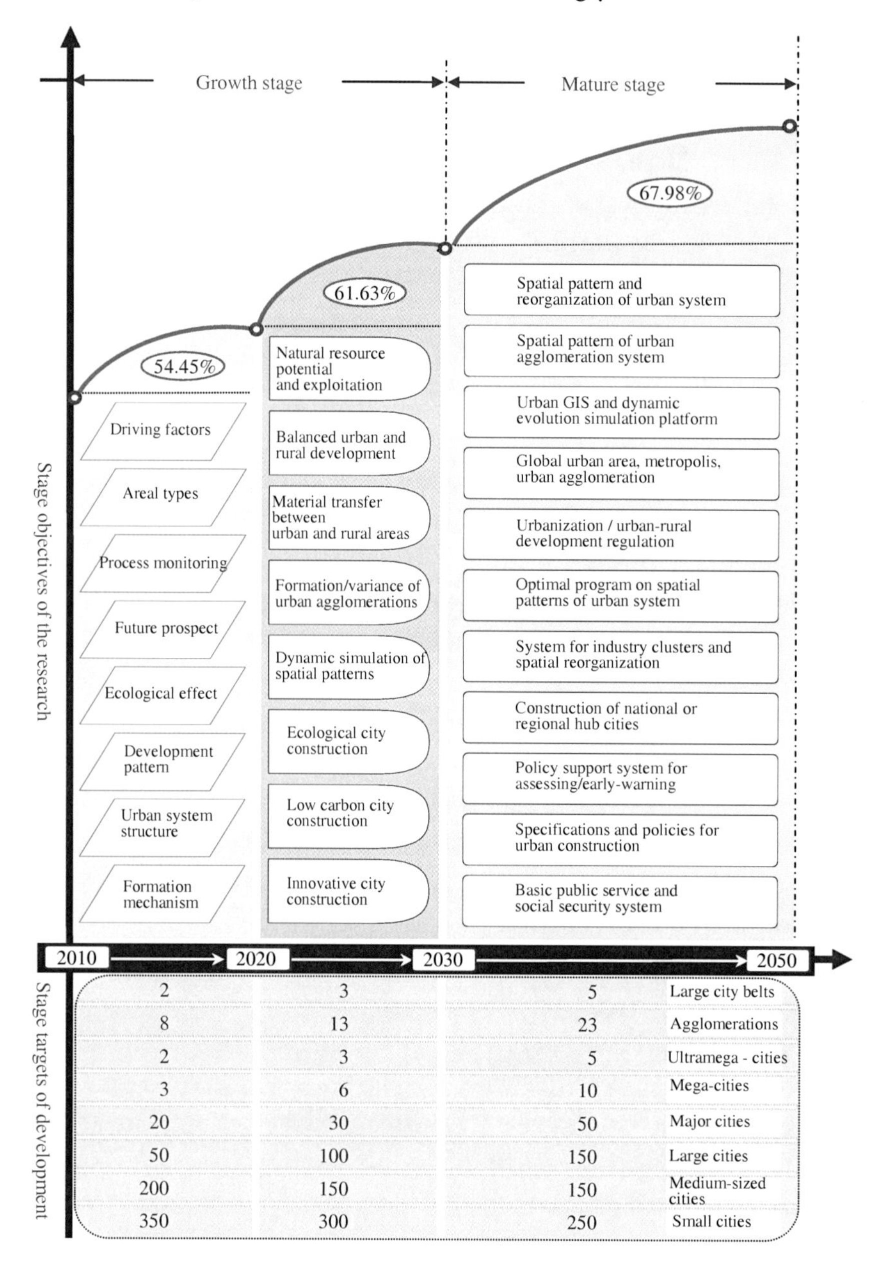

App. Fig. 1.5 Technological roadmap of urban development in China to 2005

A1.5 Leisure Demand Changes and Regional Development

We are entering the leisure era. Since the characteristics of leisure era become increasingly obvious, the changes of leisure demand and manners have been exerting great effects on society and economy, and will become new driving forces for regional development. To understand the trend of leisure demand development and its effects on space exploitation and to build up the future regional developing structure scientifically are important for promoting the coordination among regional production space, living space, ecological space and leisure space.

A1.5.1 The Status quo and Research Progress of Leisure Demand

The Athens Charter clearly addresses that recreation, along with living, work and circulation, is one of the four fundamental functions of city. It also advocates that city planners all over the world should pay due attention to the effects of recreational activities on urban space organization.

As the developed countries entering post-industrial age, and the developing countries, such as China, achieving remarkable growth in the national income, tourism industry has been developed dramatically and governments of all countries attach more importance to the increasing social leisure demand and raise the proportion of leisure economy even to a height of a strategic action for national economic and social development[137-139]. After 20 years' development, China has become a tourism economic power in the world. Under the drive of large-scale leisure demand, on the one hand, the infrastructure and social culture in urbanized regions begin to show a more obvious leisure tendency, leisure-oriented recreation centers and recreation streets emerging gradually. On the other hand, many rural areas endow their traditional agriculture with leisure, experience and recreational functions; rural tourism, ecotourism, cultural tourism and folk-custom tourism are well on the way to prosperity, and the economy of leisure tourism has been taking place of traditional industry or giving it new intension.

However, with the rapid development of tourism industry, the negative effects of high intense or over-economized recreational activities on the tourism resource, ecological environment and local community development have been increasingly concerned by researchers. Tourism activities expand into areas with fragile environment; natural and traditional cultural scenic spots are commercialized excessively; the tourism resources are developed extensively on a low level. How to rationally and scientifically standardize the development of leisure economy, promote the sustainable utilization of tourism resources, and optimize leisure space organization are new challenges for regional development in the future.

At present, researches on leisure tourism in China not only focus on tourism resources conservation, tourism products development, tourism industry development, tourists' behaviors and their recreation modes, but also extend into fields such as tourism ecological restoration, leisure space organization and social functions of leisure tourism. The current researches mainly concentrate on issues as follows:

(1) Probe into the mechanism and models of tourism resources exploitation and protection by studying sustainable utilization modes of different types of tourism resources, such as ecotourism, protective utilization of heritage sites and so on;

(2) Reveal spatial differences on leisure tourism demands and recreational behaviors to improve spatial organizing structure of regional leisure industry based on the changes of different leisure demands and recreation modes;

(3) Analyze tourism economy and its effects on the community development and ecological environment conservation.

(4) Study on the regional tourism cooperation.

A1.5.2 Changes in Leisure Demands and Their Impacts on Regional Development

Economic globalization and social diversification are changing the structure of people's consumption demands and consumer behaviors, and impelling the regional economic transition and spatial reorganization as well.

(A) Leisure-driven Economy and Recreationalization of Life[140]

Under the background of post-industrialization, regional economic development has been increasingly depending on people's confidence in leisure consumption, and their consumer behaviors and expenditures on leisure. On the one hand, people's daily leisure consumption has increased continuously. Its significant impact on the growth of the gross domestic product of all countries becomes more and more prominent. Statistics provided by the Department of Commerce of the USA shows that the current leisure expenditure of U.S. households has accounted for 26.7% of the GDP of the whole country. The data fully reflects the trend of leisurization of economy. On the other hand, the consequence of the recreationalization of life is that recreational factors are penetrating more and more deeply into work, life, shopping and other activities. The combination of recreation and traditional agriculture promotes the development of sight-seeing and agricultural leisure garden; The combination of recreation and commercial, catering industries promotes the urban transition of traditional commerce and business district so as to form a more dynamic Central Business Recreation Area; besides, there are also combinations of recreation and popular science education, recreation and body-building and health-keeping, *etc*. The recreationalization of life is penetrating into all aspects of life at an inconceivable speed.

(B) A Variety of Leisure Demands Create Vast Growth Space for Leisure Economy

The formation and development of leisure economy reflects the upgrade and transformation of industrial structure. It is not only one of the components that cause regional economic growth, but also directly contributes to the upgrade of industrial structure[141,142]. The correlation effect of leisure industry blurs the division among different industries in the economy which further strengthens the trend of industry convergence. The leisure economy has already covered a variety of industries such as tourism, entertainment, service, culture, sports, film and television industry and has even penetrated into industry and agriculture, which makes counting and analysis on these industries hard to carry out. The development of leisure economy adjusts the factors which determine the changes in industrial structure. Consequently, the effect of ultimate demands on industrial structure has been strengthened obviously, while the effect of intermediate demands is weakened.

(C) To Promote the Urban-rural Balanced Development and Spatial Pattern Transfer

The growing demands in leisure tourism will lead to new changes in the flows of people and materials between regions as well as in the regional economic structure. The intensive gear effect produced by leisure tourism activities breaks the isolation between urban and rural areas and between regions as well, which promotes the integration of economic factors, social values and culture between regions and strengthens the sense of identity, meanwhile, the interactive development of social economies between urban and rural areas is also accelerated[143]. As a result, leisure space increases largely; some of the regional land use patterns will be transfered from agricultural land or construction land into recreational land; and the recreational function is superimposed in some land on their original utilization patterns. The space of recreation activities have extended from home to outdoor, which can be distinguished as family leisure spaces and public places, and the latter includes recreational belts around metropolis, the recreational areas around large leisure holiday tourism zones and ecological leisure space based on natural ecological areas.

(D) To Promote the Conservation and Sustainable Utilization of Resources

Leisure demands will be characterized with ecology and sustainable development in the future, and the new viewpoints on leisure tourism will play a positive role in the resources conservation and sustainable utilization. With the development of leisure tourism industry, both tourism education and travel experiences will be paid more attention. Meanwhile, new concepts in tourism will be guiding the change of traditional ways of resource utilization and the protection of scarce resources.

A1.5.3 Scientific and Technological Demand in the Future

With the continuous development of economy, the rapid progress of urbanization, and the increase in people's income and leisure time, leisure has begun to play an important role in people's life in China, which brings about new scientific propositions as follows:

(1) To keep balance between leisure demands increase and regional development. How to give full play to the recreational cultural functions with the precondition of keeping harmonious social development? How to accelerate the economic development through developing leisure economy? How to promote the cultural and social progress of human being through tourism development? In a word, how to make the society more harmonious by means of leisure and tourism development?

(2) The development tendency of leisure space and its regional differentiation and rebuilding. The increase of leisure demands will make regional leisure space more systematized. Leisure space has extended from home to outdoor places and been divided into individual, family and public levels. As the leisure space becomes more and more hierarchical, we should consider the role of leisure demand and diversified recreation modes in the changing pattern of regional development. What will the structures, functions and forms of leisure space be in the future? And how will leisure behaviors and recreation modes affect the utilization efficiency of leisure spaces?

(3) Contributions of leisure economy to the growth of national economy and upgrade of industrial structure. Leisure economy, as a new type of economic activities, is production and reproduction of material goods and labor service so as to meet the demand of a special kind of consumer behavior. Since the development of leisure economy has strengthened the influence of ultimate demands while weakened that of intermediate demand on regional industrial structure adjustment, we should re-evaluate the correlation effect of leisure industry, the effects of the combination of diversified leisure activities and other industries on future leisure economy situation, and to measure the status and scale of leisure economy.

(4) The relationship between leisure activities and environment protection. As leisure tourism is becoming more and more industrialized at a large scale, leisure and tourism activities are extending to regions with vulnerable ecological environment, which has influence upon the ecological environment, traditional regional culture, and natural and cultural heritage. How will the daily changing recreation modes affect the ecological environment foundation of tourism regions? How to measure the threshold value of the environmental capacity in different tourism regions? What is dynamic management model of environmental capacity? What is the organization model of leisure space under the restriction of ecological environment?

A1.5.4 Roadmap Design

(A) Overall Objectives

(1) Analyze the trends of leisure demand and its effects on regional development.

(2) Build development models of sustainable utilization and evaluation criterion of leisure tourism resources in China; provide technological approaches, policies and measures to achieve its sustainable development.

(3)Establish simulation platform to monitor the sustainable development of leisure tourism industry.

(4)Reveal rules of recreational land distribution and factors that affect the demand and supply of recreational lands.

(B) Stage Objectives

Objectives for 2020:

(1)To explain the mechanism of interactive influences between the changes of leisure demands and regional development; to optimize the spatial organization of leisure activities.

(2)To find out a general model for the sustainable use of leisure tourism resources and establish a dynamic assessment index system; to explore technological approaches and policy measures for the sustainable use of different types of leisure tourism resources.

(3) To clarify principles of recreational land distribution and factors affecting the demand and supply of recreational lands.

Objectives for 2030:

(1) To establish simulation platform and monitor dynamically the sustainable development of leisure tourism industry by analyzing the new trends in leisure demands and regional development.

(2) To explore the time characteristics and regional differentiation of the sustainable use of different leisure tourism resources, and further improve the development models, evaluation criterion, related technological approaches and policy measures according to the social, cultural and ecological effects of tourism resources use.

(3) To adjust models and strategic approaches of recreational land use by analyzing new affecting factors of recreational land.

Objectives for 2050:

(1) To further improve the simulation platform of leisure tourism industry so as to continuously track the trends of leisure tourism demands, resource foundation, effects of social culture, effects of ecological environment and new models of sustainable development.

(2) To adjust technological approaches and policy measures of the sustainable development of leisure tourism resources according to new conditions appeared in different stages of development.

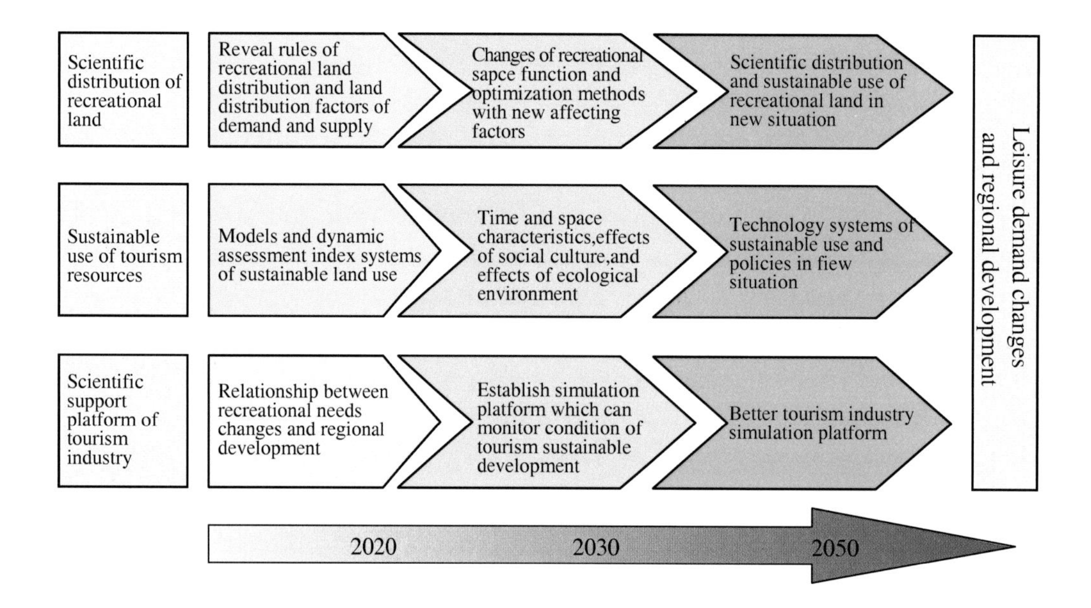

App. Fig. 1.6 Research roadmap of leisure demand changes and regional development

(C) Key Research Projects

Two key projects are to be carried out focusing on leisure demands changes and technological needs of regional development.

Project 1: Development model and sustainable utilization of leisure tourism resources

(1) To standardize scope and criterion of leisure tourism resources, and establish an identification model for the sustainable use of the resources. Analyze the internal relations between the construction of tourism facilities, modes and scale of investment and regional tourism environment, social development, and economic development. Investigate limiting factors and regulation measures of the sustainable use of leisure tourism resources.

(2) To establish an evaluation index system on the sustainable utilization of leisure tourism resources. The research objective is to find out the social, economic, cultural and environmental evaluating factors so as to set up a comprehensive index system which can effectively reflect the sustainability level of tourism resources utilization.

(3) To build up development models of sustainable use of tourism resources, including development models of sustainable use of ecological tourism resources, urban cultural tourism resources and folk custom and rural culture tourism resources.

(4) To establish monitoring platform on the sustainable development level of leisure tourism industry by taking into accounts of factors such as leisure demands, resource foundation, and environmental effects.

(5) To provide demonstration for contributions of tourism resources development to ecological protection and community development, including

comprehensive development of communities under the guidance of tourism development as well as protection of environment and traditional culture promoted by the development of tourism resources.

Project 2: Scientific distribution of leisure space and recreational lands

(1) Researches on the classification of recreational lands and measurement of the threshold value, including to establish classification system of recreational lands, calculate the per capita area of different types of recreational lands, and bring recreational land into the classification system of national land utilization.

(2) Researches on the rational development intensity of recreational lands, which consist of the measurement of the proportion of leisure construction land, the identification of the per capita area index of different types of recreational lands, *etc*. These researches are the scientific foundation of restricting and optimizing the relationship between leisure space and production, living and ecological spaces. Due to the differences in the demands for leisure land between local residents and travelers, it is necessary to probe into the relationship between recreational behaviors and the utilization intensity of leisure land. Grasp new trends in recreational development, for instance, tourism in farmlands and forests can also serve as effective complement to leisure land utilization under the conditions of meeting the increasing leisure demands and guaranteeing the major functions of farmlands and forests.

(3) Distribution and structure of recreational lands. Leisure activities of urban residents and the land value of both urban and rural areas are highly restricted by spatial location. Therefore, it is necessary to explore the issues such as the spatial location of urban and rural areas, balanced distribution of different types of recreational lands, *etc*. based on the studies of the affecting factors of leisure land utilization.

A1.6 Transport and Communication Technologies and Regional Development

A1.6.1 The Role of Transport and Communication Technologies

(A) The Improved Transport and Communication Technologies (TCTs) Play a Key Role in Reshaping Regional Development Patterns, Models, and Mechanisms

The advent of modern TCTs has profoundly affected the social, economic, and spatial behaviors of mankind. From a *spatial* perspective, TCTs are driving forces for *agglomeration* and *diffusion*, and are able to overcome barriers of physical distance and the *confined space concept* in human minds. Agglomeration results in the growth of urban systems, the formation of metropolitan areas, the centralization of population and industries, *etc*. Diffusion leads to interaction of globalization and localization, spatial differentiation of social and economic activities, *etc*. which bring changes to the elements and structures of human-nature complex systems in specific regions such as a *country* or a *region*.

Over the past three decades, the improvement and infrastructure construction of TCTs in China has been strong support to regional economic development and social progress. TCTs have: 1) played positive roles in leading the rational flow of economic elements and the spatial distribution of urban systems, optimizing regional industrial structures, and accelerating the industrial transfer among regions and cities; 2) changed the regional development patterns, models, and mechanisms in China; 3) played an important role in constructing orderly regional development patterns, optimizing spatial development systems, and cultivating hub cities and gateway cities; and 4) greatly increased the international competitiveness of the Yangtze River Delta, the greater Beijing area, and the Pearl River Delta and enhanced the location advantages and international status of hub cities such as Beijing, Shanghai, Guangzhou, Shenzhen, and Tianjin[144].

(B) China's TCT Development in the Future Will Be Characterized by Large-scale Facilities, Hierarchical Structures and Generalization of the Internet, Convenient and Just-in-time Services, and High-efficiency Management

The next three to four decades will be an important era for further development of TCTs in China. Large-scale facilities, hierarchical spatial structures and generalization of the Internet, convenient and just-in-time services, and high-efficiency management will be the important development directions. According to the analysis of a series of development conditions and trends, the planned framework of high-speed railway network and highway network will have been basically established by 2020. Meanwhile, the airline network and communication network will have reached the world's advanced level. To 2050, China's TCTs will have leapt into the front ranks of the world, and have formed advanced transport system and communication network.

From 2020 to 2030, comprehensive transport channels with large volume and high capacity will be constructed among urban agglomerations. High-speed railways, intra-city rails and magnetic levitated (maglev) railways will be under construction, and will reach a top speed of 500 km per hour. With the further modernization of communication technologies, rapid communication networks will be established among cities. The utilization of very large cargo carriers (VLCC), large-volume communication facilities, and airplanes will further promote the connection between China and the world.

These potential changes will exert important effects on regional development. Firstly, e-business, virtual economy, and logistics economy will become critical forces for regional development. Secondly, high-efficiency, intensive, and sustainable land development will be the basic development model for specific regions. This model emphasizes the rational, economical, and intensive utilization of land resources (*e.g.*, cultivated lands, coastlines, water areas, and airspace areas). Particularly, it will be possible to enhance the efficiency of resources utilization and improve the conservation of land resources by means of reducing the utilization of land resources through resources integration. Thirdly, these changes will propel further growth of international hubs and gateway cities. Fourthly, the interactive mechanism between inter-regional elements will change [38].

(C) Improved TCT Will Lead to Further Spatial Differentiation of People's Socio-economic Activities and the Formation of an Orderly and Closely Interrelated Spatial Cascade System

The *agglomeration* and *diffusion forces* which are produced by TCT improvement will be further strengthened in the next 20–30 years. As a result, socio-economic activities of mankind will be further differentiated spatially, which includes the distribution types of human activities, the intensity of spatial interaction, the development models, and the spatial patterns. The centralized distribution, convergence and integration, and coordinated development of all kinds of transport facilities networks mainly have effects on the spatial socio-economic patterns of the world, countries, or specific regions in the following two aspects: *corridor network* effects[32] and *nodal region* effects. Corridor networks depend on integrated transport and communication axes to absorb development elements together such as industries, population, and cities, and to drive and promote the socio-economic development of adjacent areas. Consequently, economic belt complexes will appear in the areas surrounding the corridors and become the most notable human-induced landscape morphology on the earth's surface which evolves following certain rules and tracks. Nodal regions follow spatial interaction and spatial economic rules, and strengthen the connection and interaction among various cities (metropolises) which are the center of human activities in the global system and country systems in order to reach a dynamic balance and form a cascade system with certain scales, rational structures, specific functions and orderly distributions.

TCT development will likely affect China's landscape morphologies in the next 20–30 years in the following ways. First, economic activities will take place mainly on major economic corridors including Harbin–Dalian, the coastline, Beijing–Shanghai, Beijing–Jiujiang, Beijing–Guangzhou, Datong–Zhanjiang, Baotou–Nanning, Suifenhe–Harbin–Manzhouli, Dandong–Baotou, Qingdao–Taiyuan–Zhongwei, Longhai (Lianyungang–Lanzhou), Nanjing–Xi'an, the Yangtze River, and Shanghai–Kunming, *etc*. Especially the industry and economy belts along the coastline and major rivers will become China's first-class axes for land development, which will dramatically change the natural land structure and the patterns and mechanisms of the man-earth areal system in China.

Secondly, transport economic circles will appear gradually. In this process, industries and population will further agglomerate or disperse according to the principle of a pole-axis system; economic and social spatial organization will be optimized through industrial connection mechanisms and social relations. Hence the transport economic circles will be formed, including half-an-hour, one-hour, or two-hour economic circle, *etc*., which will direct the orderly evolution of spatial structure. Transport economic circles will probably be formed in some urban compact districts in China such as in the greater Beijing area, the Yangtze River Delta, the Pearl River Delta, the center and south of Liaoning province, Shandong peninsula, the Shangsha–Zhuzhou–Xiangtan area, Wuhan, the great Zhengzhou area(central plain), the Chengdu–Chongqing area, Guanzhong Plain and the Changchun–Jilin area. Thirdly, some of the cities mentioned above will grow into international gateway cities. The expansion, crossover, and integration of transport networks such as railways, air transportation, and waterway will enhance the accessibility and location advantages of some megacities, and promote them to become national gateway cities, or even the hubs of the Asia-Pacific Region. Shanghai, Beijing, Guangzhou, Tianjin, Shenzhen, and other major cities will probably be on the list. These cities are not only the throat for the international communication of China, but also the hubs of socio-economic exchange among regions[145–149].

A1.6.2 Major Research Objectives

(1) The relationship between TCT improvement and new economic patterns. E-business, virtual economies, and logistic economies, the formation and development of which are closely related to TCT improvement, will become important forces in regional development in the future. TCTs are the major factors in shaping new economic patterns. Therefore, to study their working mechanisms, intensity, and spatial-economic rules will continue to be one of the major tasks in the future.

(2) The change of the working mechanisms of regional development elements. The advent and development of modern transport has overwhelmingly changed the utilization of elements and their proportions in a country or a region, the spatio-temporal relationship of their interaction, and the logical-balance relationship between them. Closely tracking the improvement of TCTs and dynamically studying the relationships stated above will help us understand the world accurately and rationally design self-development models and paths.

(3) The development model and mechanism of high-efficiency and functional lands, and the role, intensity, and effects of TCT improvement. A series of scientific studies are required to seek precise thresholds or parameters.

(4) The human environmental effects of the integrated transport system development. From the comprehensive perspectives of ecology, society, and economy, we can explore the relationship between the improvement of transportation technology and urbanization, the *synergetic* process and *coupling* relations in regional development, and the *synergetic* and *non-synergetic* spatial effects and efficiency; study the effects of highway networks on the social economy and resources environment in China; examine the

social and environmental effects of the infrastructures in major metropolitan areas; and analyze the scientific basis of *compact city* development under the direction of high-efficiency land use.

(5) The human mechanism of material flows and their effects on the evolution of human-nature complex system. From the perspective of geography, we can study the spatial and economic significances of TCT improvement and the development of infrastructure service networks; deepen the related theories about spatial economic location and spatial organization of industries; explore the mechanism of spatial interaction and the generation mechanism and evolution laws of spatial flows (*e.g.*, material flows, population flows, information flows, and energy flows); search for scientific parameters of general significances so as to provide support for regional, metropolitan and ecologic planning. Also, we can examine the relationships between the space of flows and the improvement of communication technologies, logistics systems and TCT improvement, and e-business and its transport communication foundations.

A1.6.3 Roadmap Design

(A) Stage Objectives

(1) Objectives and tasks for 2010–2020. The coming ten years is a key era for strengthening the studies on TCTs and identifying further scientific propositions. In view of the current research progress of TCTs and practical research needs in this field in China, we should concentrate on studying the inner mechanisms of transportation technology improvement and regional development, revealing their relationships, and exploring their general rules in this stage. To be specific, the researches will include the following subjects: the interactive mechanisms between the improvement of transportation technology and regional development, the relationship between development patterns of high-efficiency and functional lands and transport infrastructure networks, and the evolution laws and dynamic mechanisms of human-nature systems in specific regions.

(2) Objectives and tasks for 2020–2050. These thirty years is the medium- and long-term research stage which is a major one for the in-depth researches in this field. According to the research objectives in the period of 2010–2020 and future research needs, we should analyze the expansion of the integrated transport network, investigate the effects of spatial flows on nature, mankind, and physical environment, comprehensively examine the spatial effects of transport infrastructure development, and adjust and optimize the inner mechanisms and regulation methods of transport infrastructure networks and regional development in this stage. To be specific, the researches will include the following subjects: the synergetic process—coupling relationship between the improvement in transport technology and urbanization, regional development, the optimization mechanisms and regulation methods of transport infrastructure networks and regional development, the economic and resource-

environmental effects of highway networks, the social and environmental effects of infrastructures in metropolitan areas, and the relationship mechanisms of the space of flows and TCT improvement.

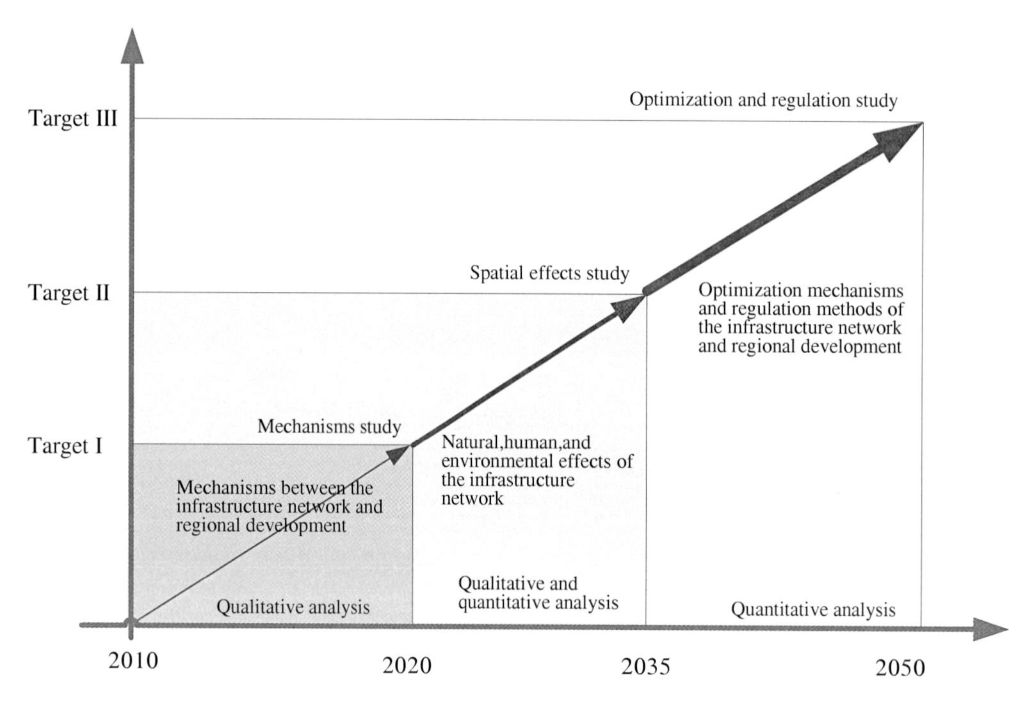

App. Fig. 1.7 Research roadmap of transport infrastructure networks

(B) Major Tasks

a. Relationship between Transport Infrastructure Networks and Spatial Development

Infrastructures are not only an important component of the spatial structure, but also a major element which is considerably significant to optimize the spatial organization of regional development. Therefore, to scientifically understand the relationship between infrastructure networks and spatial development will be a scientific research proposition in this field in the future.

Researches in the following aspects should be strengthened in the future on the basis of existing research issues:

(1) Relationship between infrastructure networks and regional development;

(2) Relationship between the development of high-efficiency, functional lands and infrastructure networks;

(3) Coupling relations and interactive mechanisms between the expansion of infrastructure networks and the evolution of human-nature systems in specific regions.

In doing so, a key scientific task is to be solved–the spatial-temporal and logical-balanced mechanism of infrastructure networks and regional development. Systemic inspection on the general relationship mechanism of infrastructure networks and regional development can be conducted through related researches, which is of considerable scientific supporting significance to the regional development in China .

b. Influences of Transport Infrastructure Networks on Nature, Mankind, Society, and Physical Environment

The construction and improvement of infrastructures have changed the natural and social conditions as well as the economic foundations in some regions, and have guided the moving directions of material and information flows, which greatly affect the status quo and future trends of regional development. Thus, an in-depth study on the comprehensive effects of infrastructure networks on nature, mankind, society, and physical environment and their mechanisms, and an investigation on the adjustment and optimization of the relationship mechanisms and regulation methods of infrastructure networks and regional development will be key research targets in this field in the future.

The following subjects should be emphasized in our future researches:

(1) Synergetic process and coupling relations of transport technology improvement, urbanization, and regional development;

(2) Economic and resource-environmental effects of highway networks;

(3) Social and environmental effects of the infrastructures in metropolitan areas;

(4) Relationship mechanisms of space of flows and TCT improvement.

In order to carry out the above researches, another key, scientific task is to be solved—the effects of infrastructure networks on nature, mankind, society, and the environment. We will make effects to fully reveal these effects through related researches in some regions, to probe into the relationship mechanisms of infrastructures and regional development from a higher level, to discuss the adjustment and optimization mechanisms and regulation methods of infrastructure networks and regional development, and thus to provide important scientific support for China's regional development and infrastructure construction.

A1.7 The Status Quo and Research Progress of Ecological Environment and Resource System in China

A1.7.1 The Research Status Quo of China's Ecological Environment and Resource System

(A) The Status Quo of China's Ecological Environment and Resource System

According to many scholars' empirical researches, China's ecological footprints show a fluctuating but increasing tendency, the ecological carrying capacity has been on obvious decline[150], and China's ecological environment as a whole fell into a serious situation. The ecological carrying capacity shows significant spatial differences at the provincial level: eastern regions except Hainan, Guangxi and Fujian, have experienced rapid decrease; both decline and increase are seen in central China. The ecological carrying capacity of Tibet, Sichuan, Yunnan and Guizhou has been enhanced greatly while that of Qinghai, Gansu, Ningxia and Shaanxi has descended. So far as ecological fragility is concerned, the situation in central and western China is severe , and continuous areas with high and relatively high fragility have been formed to the north of the Hu Population Line (Hu-line), including the Tibet-Qinghai Plateau, the humid and subhumid regions of Loess Plateau, the temperate zone in Inner Mongolia Plateau and the desert area in the northeast temperate and warm temperate zone.

"Abundant total but low per capita amount" is the shared characteristic of resources in China. Since 1996, the total quantity of energy resources supply and demand has transformed from net export into net import, the energy resources utilization efficiency has increased rapidly but still much lower than developed countries, and the coal-based energy consumption structure is still difficult to change. The total water consumption of China increased rapidly with an average annual growth rate of approximately 10% from 1949 to 1997, and showed a slight fluctuation between 530 billion to 580 billion cubic meters after 1997, which accounts for 70%–80% of the usable water resources in China. As for the spatial distribution of water resources, the conflict between water supply and water demand is conspicuous in the water basins of north China[151], while drainage basins in south China have relatively rich water resources but have suffered the water quality-induced water shortage in recent years. The cultivated land areas in China increased during the period of growth from 1949 to 1957, and showed a decreasing tendency afterwards. At present, the cultivated land per capita in China is only 0.1 hm^2 and the reserved land resources for cultivation are scarce. And ecological de-farming and the occupation of cultivated land for urban construction use are the main causes of cultivated land loss of central-western China and eastern China respectively[152].

(B) The Research Progress of Ecological Environment and Resource System

The researches on ecological environment and resources system are getting more and more attention as people critically reflected the traditional economic growth mode. After the introduction of the concept of sustainable development, scholars in the fields such as economics, geography, environment, resource, ecology, *etc*. began to concern more about the interaction mechanism between natural ecological system and socio-economic system and their evolutionary laws in order to provide theoretical guidance for the practice of sustainable development in China. At present, the researches in the field of ecological environment and resources system in China mainly focus on the evolution and risk pre-warning and evaluation of ecological environment and resources system, the interaction between the resource environmental system and urbanization and industrialization, the prediction of ecological environment and resources system development, and the development paths and experiences of developed countries, *etc.*[153-165]

Ecological carrying capacity is the mainstream method for evaluating the status of ecological environment system. Empirical studies have been widely carried out on national, provincial and urban levels, but difference in research methods, statistical standards and some other problems still need to be solved. The studies on the evaluation of resource system development still focus on the balance between aggregate supply and demand and structural evolution. Under the background of the long-term condensation model economic development, urbanization and industrialization have been considered as the main affecting factors of the ecological environment and resources system of China. And the influences of the rapid, extended and heavy industrialization and the unpractical urbanization development mode on the ecological environment and resources system have been extensively studied.

A1.7.2 Judging the Future Development of Ecological Environment and Resource System and the Scientific and Technological Demand Analysis

(A) Judging Ecological Environment and Resource System Development

a. Judging Ecological Environment Development

Timely and appropriate laws and regulations as well as policies are the important driving force for promoting the improvement and restoration of ecological environment system, but different degrees of strictness of the policy implementation will exert different influences on ecological environment. In view of the Outline of the 11th Five-Year Plan of Ecological Environment of China and the implementation of "Major Function Oriented Zoning" Strategies, in loose policy environments, the ecological environment of China will continue deteriorating before 2020 and will not be improved until 2030, but the environmental degradation will be contained in 2020 and an obvious improvement will have been achieved by 2030 if under strict policy conditions.

b. Judging the Resources System Development

Along with the sustained and rapid economic growth, the supply capacity of resource system will be faced with more and more pressure. Judging from the present trend of resource supply and demand and energy utilization efficiency, China's dependence on foreign energy will increase continuously. Under positive conditions, China will remain self-sufficient in coal consumption, but its demand of petrol and natural gas will reach 0.22 billion t and 60 billion m^3, with the degrees of external dependence being 55% and 33% respectively in 2020; under negative conditions, China will have a coal supply gap of 1 billion t and a natural gas supply gap of 100 billion m^3 in 2020 [155,156].

According to previous predictions of water demand and the trend of water use in developed countries and the economic development objectives and urbanization trend of China, if strict water use and water-saving policies are implemented, zero increase in industrial water use will be achieved, the total amount of industrial water use will begin to decrease substantially, the inflection point of which will appear, and the peak value of the total amount of water use will be about 650 billion cubic meters in 2020; however, if under relaxed policies, the zero increase in industrial water use and total amount of water use will not be realized until 2030, the peak value of total amount of water use will be about 700 billion cubic meters, which accounts for 85% of the usable water resources, and the lowest point of per capita water resources will be 1700 cubic meters. Thus, without the implementation of strict water use and water-saving policies, China will soon be confronted with water resources crisis.

If current development trend continues, the cultivated land resource system in China will be faced with huge challenges. If the actual cultivated land area changes as shown in the following table, the cultivated land tenure red line of 0.12 billion hm^2 will be broken. And before 2050, China's cultivated land resources will be able to meet the demand for crop planting[166] but fail to satisfy the cultivated land demand under multi-objective demand schemes[167].

App. Table 1.2 The actual cultivated land area under the prediction of different schemes

Year	2020	2030	2050
The actual cultivated land area/10^9hm^2	0.116	0.116	0.115

Note: Prediction method: Before 2010, it is calculated according to the average annual net decreasing area from 2001; the ecological de-farming will be basically finished in 2010. From 2010 to 2020, the average annual decrease of de-farming will be 100 per year, and the rest are calculated according to the annual average value from 2001; after 2020, the ecological de-farming will be 50, and urban occupation will be 50% of the annual average from 2001 due to the counter-urbanization.

(B) The Scientific and Technological Demand Analysis

The factors causing the ecological system deterioration in China are multiple, including the lack of knowledge on the operation mechanism of eco-environmental system, the weak awareness of eco-environment protection, the unreasonable protection and restoration means, *etc*. Therefore, we must develop innovations in natural sciences and social sciences, popularize ecological environment protection knowledge, and enhance the evaluation capacity of ecological environment in economic development and the ecological environment monitoring, protection and restoration capacities based on systematical understanding of the operation mechanism of eco-environmental system and the interaction mechanism between eco-environmental system and socio-economic system.

In order to ease and coordinate the relationship between the resource system and industrialization and urbanization, and thus avoid the crisis of resource system overload, scientific and technological innovations and appropriate development modes of industrialization and urbanization should be developed. It is necessary to save the energy and resource consumption per output unit by enhancing the utilization efficiency of energy and resources, to extend the width and depth of resource exploitation and utilization by developing the technologies for the exploitation of new energy and renewable resources, to increase the total amount of energy and resource supply through waste recycling technology and the energy cascade utilization technology, to reduce the waste of energy and resources through technological transformation of traditional industries and developing high-tech industries, and to reduce the disturbance and destruction of natural ecological system caused by urban development and balance the material, energy and information exchanges between urban system and natural ecological system through the innovation and popularization of scientific technologies such as urban planning, green architecture, and rainwater harvesting and utilization, *etc*.

A1.7.3 The Overall Goals and Stage Objectives to 2050

The sound development of ecological environment and resource system is the significant foundation of sustainable socio-economic development and the important content of comprehensive sustainable development. In order to ensure the smooth achievement of the socio-economic development targets by the mid-21st century, we need to deeply implement the ecological civilization strategy, strictly carry out the outline of the 11th Five-Year planning of ecological environment, "Major Function Oriented Zoning" policies and new energy industry development strategies, develop social economy within the carrying capacity of eco-environmental and resource system, enhance our understanding on the internal laws of the evolution of eco-environmental and resource system, coordinate the relationship between socio-economic system and eco-environmental and resource system, put forward the improvement and restoration of eco-environmental system and the enhancement of the carrying capacity of resource system, and strengthen the sustainable development capacity in China through scientific and technological innovation and timely adjustment of development modes.

By 2020, the deterioration trend of ecological environment will have been slowed down and effectively contained in accordance with the development goals of realizing an overall well-off society, and the carrying capability pressure of resource system will have been relieved. Ecological and environmental protection policies will be effectively implemented, the eco-environmental effects of socio-economic development will be significantly improved, and the ecological environment of some regions will turn for the better. The coal resources will be more than self-sufficient, the external dependence degree of oil and gas resources will be controlled within 30% and their import demand, will be well guaranteed, and the proportion of new energy utilization will increase greatly. The inflection point of total water and industrial water use will appear, and the peak value of total water consumption will be controlled below 650 billion m^3. The level of intensive utilization of land resources will be enhanced, and the total amount of cultivated land resources will be maintained above 120 million hm^2. Large-scale circular economy will be achieved, the ecological footprint and resource consumption per unit output will decline, and the gap between China and developed countries will be significantly narrowed.

By 2030, the ecological environment will have been significantly improved, and resource system will be able to effectively meet the demand of energy and resource. The historical transformation of ecological environment protection will be basically accomplished, the coordinated development of ecological environment and social economy will be achieved, and the ecological environment in most of the regions of China will show an improved tendency. The energy utilization structure will change greatly, the proportion of new energy and renewable energy utilization will reach 30%, and the energy-saving effect will be very obvious. Total amount of water consumption will begin to decline, the unreasonable spatial allocation of water resources will be improved, and the water quality-induced water shortage will be alleviated. The cultivated land resource will maintain at 120 million hm^2.

By 2050, China's environment and resource system will have entered a new stage of virtuous cycle. The eco-environmental priority strategy will be strictly implemented in industrial and urban development planning, the eco-environment system will be highly stable, and the sustainable development capacity will be continually enhanced. The proportion of new energy and renewable energy will be above 50%, and fossil energy technology will be popularized. The total amount of water use will drop to 500 billion m^3, and the water use efficiency in China will reach the level of developed countries. Total amount of cultivated land resources will be able to meet the multi-objective demand.

A1.7.4 Implementation Plans and Technological Roadmap

In order to promote the sustainable development and realize the coordination between socio-economic development and eco-environmental and resource system, we must put forward socio-economic development modes in accordance with the national conditions in China on the basis of the understanding of the virtuous circle mechanisms between natural system and socio-economic system, move away from the irrational development modes under which economic growth is over-emphasized while eco-environmental effects and the support capacity of resource system is ignored, emphatically study the interaction mechanism among ecological environment, resource system and socio-economic system, the evaluation methods of the state of ecological environment and resource system, and the development modes and optimal regulation on social economy under the restriction of natural systems.

(1) To develop innovative and unified scientific evaluation and prediction method of the development state of ecological environment and resources system.

At present, there is no shortage of relevant evaluation methods, but the lack of a standardized one for empirical studies makes it difficult to carry out effective horizontal and vertical comparisons, which in turn hinders the revealing of the state and evolutionary laws of the eco-environmental and resource systems in China. Therefore, we must proceed from the national conditions in China, study the relevant evaluation methods, summarize the key affecting and characterization factors, develop innovative and unified evaluation and prediction method of the development state of ecological environment and important resource subsystems such as energy, water resources, land resources, *etc*, and reveal objectively their evolutionary trajectories and their capacity to support the socio-economic development in China.

(2) To reveal the response mechanism among ecological environment, resource system, industrialization and urbanization.

Based on the domestic and foreign research achievements in the interaction mechanism between natural and socio-economic system, integrating economics, management science, systematic science, ecology, geography and other related disciplines, combined with the comprehensive theoretical study and the analysis on the impact of key factors, construct the interaction models between ecological environment and industrialization, ecological environment and urbanization, resource system and industrialization, and between resource system and urbanization. Analyze the eco-environmental effects and reaction paths and modes of industrialization and urbanization in different regions and development stages, and discover the response mechanism of the industrialization and urbanization to ecological environment and resources system under the national conditions in China through national and regional empirical studies.

(3) To put forward the development strategies of urbanization and industrialization under the restriction of ecological environment and resources. Based on the response mechanism of ecological environment, resources system, industrialization, and urbanization, we will develop the simulation and prediction models for industrialization and urbanization development under the restriction of ecological environment and key resources through summarizing and integrating the affecting factors of the natural, social and economic development in China. Furthermore, taking sustainable development as the goal orientation, we will put forward development strategies of urbanization and industrialization in accordance with the national conditions in China, and determine optimization and regulation approaches and measures of the development modes of industrialization and urbanization at different development stages.

It is a complicated and systematic project to promote the coordinated sustainable development of ecological environment, resources system and socio-economic development for the reason that both the academic research innovations in the interaction mechanism among ecological environment, resources system, industrialization and urbanization and the scientific and technological innovations in ecological and environmental protection and restoration and technologies for efficient resource utilization need to be put forward in accordance with the different national conditions and development goals of different stages. Corresponding technical roadmap is shown in App. Fig.1.8.

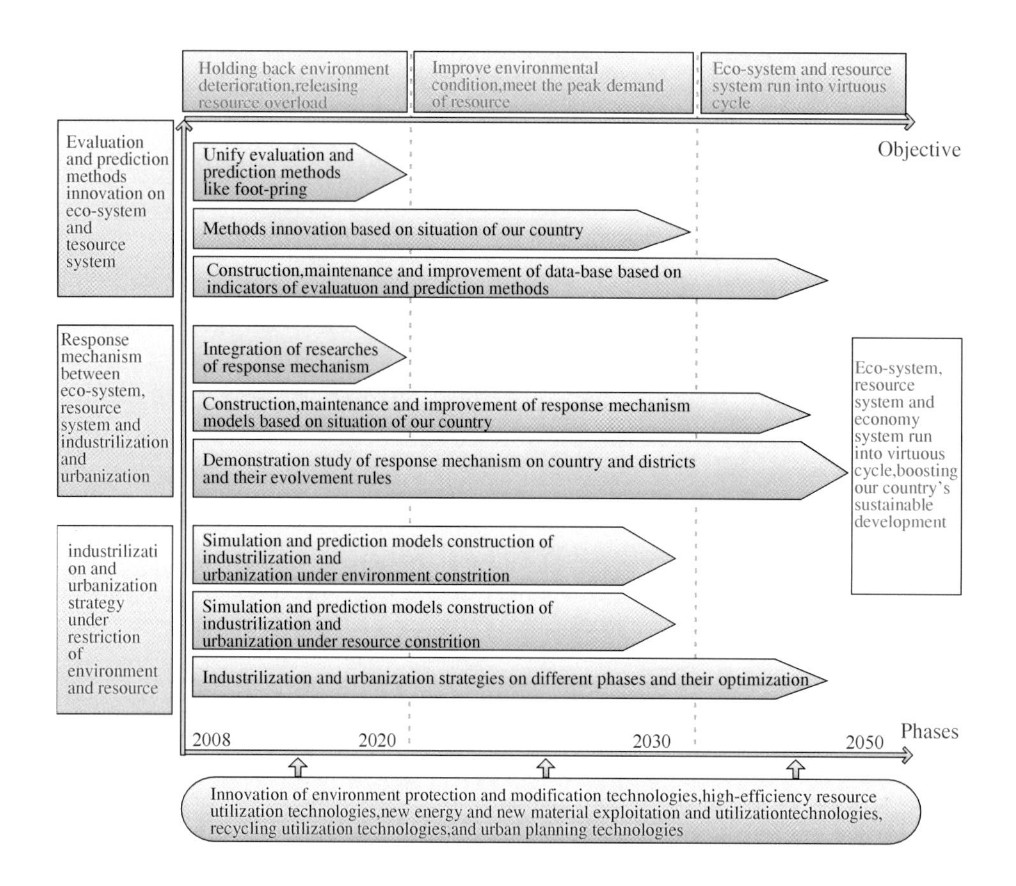

App. Fig.1.8 The technological roadmap for the development of ecological environment and resources system

A1.8 Roadmap of Regional Management Systems

A1.8.1 The Status Quo and Research Progress of Regional Management

In China, the focus of regional development strategies is shifting to the construction of the orderly development of spatial structures, which will further enhance the demand for effective regional management. The practices and experiences of western countries as well as China demonstrated that the reforms in regional development systems had been driven by marketization, industrialization, urbanization and modernization. This process was greatly prompted by the social movements of local autonomy and public-involvement[168], and supported by changes in social values and value orientation[169]. The administrative reform of governments had provided strong institutional basis for the transformation of the regional management systems[170]. Based on the understanding of the influencing mechanisms of these factors and general judgment on the development trends of regional management, it is decided that the core content of the roadmap design on the development of regional management systems should be the effective regulation on the policy elements.

The experiences of western countries show that under the influence of an increasingly globalizating economy and regional competition, regional management systems are becoming increasingly polarized. On the one hand, international and regional organizations are taking over the powers of individual countries through various international free trade agreements, common markets, and economic cooperative unions such as the EU. On the other hand, the administrative powers are deconcentrated to local authorities such as metropolitan governments since they become more flexible and economically influential in global competitions. In addition, regional management has been increasingly emphasizing more on the involvement of multi-participants and the democratization and legislation in decision-making[171], more on regional-based public policies rather than those within administrative districts, more on the legalization and informatization of its measures[172], more on the precision rather than the simplicity of the tools of management, and more on the diversification in its objectives which covers management in economic, social, environmental and infrastructural aspects.

Under this background, the conventional functions and positions of the governments have changed a lot. A phenomenal change is that the government services are becoming more customer-and market-oriented. Competition mechanisms are introduced into the field of public services, and the decentralized methods of management are widely adopted in order to smash the monopoly of governments in providing public services. Analysis on cost and efficiency and the effect and quality of public services are stressed. Strict executive regulations are relaxed and replaced by the control of performance objectives. Through the reform of the administrative systems, the government capacities have been enhanced and its governance has become more effective.

Based on the review of western experiences and the regional development in China, the Urban and Regional Management Research Team of Chinese Association of Geography (2008) proposes that four aspects have to be concerned on at present in China. First, problems in the economic development and national economic security brought by the changes of global economic environment, regional industrial division and industrial transfer should be paid much attention to. Secondly, it is urgent to organize and expand regional urban agglomerations in order to adapt to the rapid changes of international situations, for which the governments must break down the borders of administration, to foster the innovation hub cities leading the economic development of the whole region, and to promote the transfer of population. Thirdly, urban growth management is presently an important direction for the urban and regional management in China which includes the management of urban environment development, urban land use, and urban employment. Fourthly, the management of regional knowledge is greatly needed, the studies on which is yet far from adequate now in China.

A1.8.2 Development Trends and Scientific and Technological Demand

From now on, the regional governance of Chinese government will be strengthened through the "Major Function Oriented Zoning" spatial governance. The "Major Function Oriented Zoning" provides important basis for making and implementing diversified regional management policies [13]. The regional management systems must be improved in order to put the each zone's functional attributes into practice, to strengthen the division of labor and cooperation between different function zones, and to ensure the fairness of regional development. First, the legal system of spatial governance should be immediately improved to clearly define the specific duties of governments at different levels and the relationship between governments and the people. Secondly, the ecological conservation and ecological compensation must be taken as important means of the macro-control of the government. In addition, the consistency between the policy target areas and the "major function zones" should be strengthened in order to ensure effective regional governance.

In the next 10–20 years, the high-speed economic growth in China will continue, thus the conflicts between socio-economic development and the natural bases are quite likely to be intensified. In order to keep the ecologically vulnerable areas from being deteriorated and to improve the sustainable development capacity of urban agglomeration areas, issues of territory resources and ecological security ought to be studied in depth.

Along with the progress of globalization, the flows of capital and labor are continuously accelerated. As the formation of the global urban system, the competitions among urban agglomerations become more and more intensive. To adapt to the new situation, the competition mechanism should be introduced into the government management system to improve the effect and quality of public service. Besides, the positioning of cities in the global economic network should be re-defined so as to form competitive advantages in global economy. At the same time, the development of powers and structures as well as the division of labor and relationship between the multiple participants of urban agglomerations should also be re-studied.

In order to meet the above requirements, the following issues should be emphasized in the researches on regional management systems:

(A) The Overall Spatial Patterns Strategies of Territory Development and Effective Utilization of Spatial Resources

Although China is a country with vast territory and abundant in energies and resources, the resources per capita are far from adequate due to its huge population. In the past decade, the economic and social development was accompanied by extensive occupation and land damage, consumption of water resources, pollution of aqua-environment, and excessive consumption of energies and mineral resources. If these trends continue in ten years, there will be no land to use in some cities; the arable land per capita in several densely populated provinces will decrease to 0.3 *mu*; the food self-sufficiency rate of the economically developed eastern coastal areas of China, where 60% of the national population dwell, will drop below 70%. At the same time, the continuous import expansion of energies and resources will also lead to greater international political pressure and conflicts. In order to meet the challenges in homeland security, the spatial development pattern across the whole country should be reshaped, where the development and construction have to be effectively planned and regulated.

(B) Gap and Imbalance of Regional Development

The differences in natural and resources basis, historical conditions, locations, innovations in science and technology, and institutional backgrounds result in regional development gap. In recent years, the development of globalization and informatization has also widened the gap. As these factors remain, the imbalance in regional levels and spatial structures will be a long-term tendency, which requires the governments to make constant choices and adjustment between balance and unbalance. Thus it is necessary to carry out both theoretical and empirical researches on the processes and mechanisms in which development gap among different regions is widened or narrowed.

(C) Mechanism of Regional Development Under the Three-dimensional Purposes of Economic Growth, Social Welfare and Environmental Preservation

The regional development will require not only the realization of economic growth, but also social welfare and environmental preservation. Under the three-dimensional objectives, the process control and spatial governance policies should be adjusted, and new economic growth theories and analytical methods should be developed. Specifically, the interactions between the three-dimensional objectives should be clarified, the process, characteristics and critical factors of economic development, which are different from those under the single-objective system, should be revealed; at the same time, the process, structures, and governance policies of regional development should be simulated and analyzed.

(D) Tools and Means for the Governments to Conduct Spatial Governance

It has been commonly recognized that spatial planning is essential for the governments to conduct macro-regulations under the market economy system. The governments intervene in public fields mainly through regulations of resources, especially financial funds, lands, and other resources so as to guide, restrict, or encourage the activities of the private sectors and strengthen public involvement in the implementation of plans[173-175]. In such a situation, the following scientific problems are to be solved: the types, amounts and changes of the resources that are under the regulation and control of the governments; the intensity of regulation and working mechanisms of different types of resources; ways in which enterprises and the public can be guided through governments' regulation on resources.

(E) Cross-border Regional Cooperation

In recent years, the cooperation between China and its neighboring countries and areas has been increasingly deepened and expanded, which should be taking into account seriously in the public policy-making. The cross-border regional economic cooperation entities include the Northeastern Asian Cooperation Region, the Central Asian Cooperation Region, the Lancang River-Mekong River Cooperation Region, the Sino-Indian-Nepal Cooperation Region, the Eastern Asian and Southeastern Asian Cooperation Region, and the Sino-Australian-New Zealand Cooperation Region, *etc*. In order to cope with the geo-political and geo-economical cooperation and conflicts, prospective studies should be conducted on the economic cooperation foundations, cooperation frameworks, selection of economic centers, and large-scale regional infrastructures in the cooperation regions mentioned above. The energy issues and regulation frameworks of global environmental cooperation, as the major concern of cross-border regional cooperation, should be scrutinized in particular.

(F) Regional Management for New Economic and Emergent Events

The global movement of *greenhouse gas emission reduction* brings about new opportunities and challenges to the adjustment and upgrade of regional industrial structures. Especially the regions based on a low-carbon industrial structure will take more advantages in the new trend, and the spatial pattern of the foreign trade in China will also be reshaped. In such a situation, prospective studies on the trading system, price formation mechanism and market situation of the low-carbon economy should be conducted. In addition, due to the complexity of the socio-economic systems, prospective studies on emergency mechanism and plan of regional emergent events and the precautions for recovery have become more and more critical to reduce the losses caused by unpredictable emergencies such as natural disasters, epidemic diseases, and terrorist attacks, *etc*.

(G) New Regional Management Policies

To ensure an equal access to the basic public services of all people, it is necessary to refine the public investment policies and the standard of the allocation of public facilities. An effective cross-regional mechanism for ecological compensation should be established so as to deal with related problems such as estimation on the gains and losses, the ecological compensation standard, and the subjects and methods of compensation. New system for assessing political achievements should be established in which ecological and environmental indices are incorporated.

(H) Monitoring, Forecasting and Pre-warning Systems for Regional Development

A national-wide standardized system of regional division should be established so as to provide scientific basis for the evaluation on the status quo of regional development and the implementation of regional planning policies. It is also necessary to track the new trends in urbanization and environmental development, to identify the new factors in regional development and to guide the economic adjustment and spatial restructuring through monitoring the progress of regional development dynamically. Last but not least, a pre-warning system should be set up in order to keep the imbalance within a controllable range and to avoid social turmoil.

A1.8.3 Overall Goals and Stage Objectives to 2050

By 2050, China will have reached the level of a medium-developed country in the world, modernization will have been basically realized, China will have become the largest economic entity in the world, and the percentage of urban population will have increased from the present 43% to 65%–70%. This is a critical period of adjustment for China in which the processes of industrialization, urbanization and modernization will experience a transition from high-speed growth to stable development, the total population will reach the peak and then decrease gradually, and the spatial structure of socio-economic development will no longer conflict as before but tend to be balanced instead. These bring about great challenges to regional management.

(1) To 2020, the total population in China is going to reach 1.43 billion, and the urban population will be about 0.79 billion which accounts for nearly 55% of the total number. The GDP will amount to 47 trillion RMB, and the per capita GDP is expected to increase to 33 thousand RMB. The predicted industrial structure is approximately 11 ∶ 50 ∶ 39.

To 2020, the Yangze River delta (including Shanghai, Jiangsu and Zhejiang provinces), Guangdong province and Beijing-Tianjin metropolitan regions in the coastal areas will be the main target regions of the migration of population in China which are mainly from the middle and western provinces adjacent to these areas. So far as each province is concerned, the migration from rural to urban areas will increase. As a result, the speed of urbanization will be extremely high, and regional development or even the overall social economy will be increasingly restricted by natural resources such as energy, land and water resources as well as environmental and ecological conditions. In such a situation, the trend of environmental deterioration will be very difficult to change completely. Nonetheless, it is expected that, with the establishment of the "Major Function Oriented Zoning" across the country, a more effective, scientific and standardized system for regional management will be built up, and the market mechanism, cooperation mechanism, compensation mechanism as well as the supporting mechanism will also be improved then. At that time, the refined regional management will emphasize more on social justice, environmental preservation and other non-economic objectives. And on the basis of the "Layout of major function zones", the problematic regions such as natural resource-based cities, undeveloped areas, and ecologically degraded areas will get more support from the government.

(2) To 2030, the total population in China will reach its peak of 1.56 billion, and urban population about 0.95 billion taking about 60% of the total number. The GDP is expected to be 80 trillion RMB, and the per capita GDP will be 63 thousand RMB. The predicted industrial structure is approximately 8 ∶ 47 ∶ 45.

From 2020 to 2030, the speed of urbanization will slightly slow down with an annual increase of 0.5%–0.6% on average, which is lower than the 0.6%–0.8% during the period of 2010–2020. However, it still means that about 16 million peasants will become urban residents each year, which is a huge burden to the society. Due to the huge population, the natural resources by per capita will become more and more deficient, the security problems of energy and food provision will be prominent, and the conflicts between development and preservation will be intensified. If the objectives of resource-saving and environment-friendly society can be realized, the pressure of resources, environments and eco-systems will be released. Instead, how to deal with the internal problems in each region will be the main task, which calls for smart and accurate regional management more than ever. As the population begins to decrease after 2030, new social and economic problems such as aging population, the labor loss, social welfare, collapse of rural areas, *etc*. might occur. These problems will become new challenges to regional management.

(3) To 2050, the total population in China will be 1.38 billion, with about 1.0 billion urban population, which accounts for about 65%–70% of the total number. The GDP is expected to be 150 trillion RMB, and the per capita GDP to be 123 thousand RMB. The predicted industrial structure is approximately 6 ∶ 42 ∶ 52.

With about 400 million peasants becoming urban residents and the urbanization rate reaching 65%–70%, the large-scale and rapid urbanization process will almost be completed in the following 40 years. The construction of the resource-saving society will be generally accomplished and the environmental and ecological pressures will be significantly relieved. According to the inverted U-shaped relationship between regional imbalance and economic growth, the spatial structure of social economy in China will gradually reach a balanced state. However, as the total amount of economy is huge, the tasks of spatial organization will be more complicated. Under such a background, regional management will pay more attention to the construction of social welfare systems; the gaps between urban and rural areas, the gaps between different regions, and the gaps between different social classes within the same region will be greatly narrowed; as 65%–70% of the population living in urban areas, the total amount of energy consumption will double the present level at least. The regional management will concern more about resource support capability and eco-environmental problems caused by resource consumption; Due to the limitation of land and resources, competition for new types of resources such as the ocean and outer space resources will be more intensive, which will also become the important contents of the regional management and regional cooperation.

A1.8.4 Implementation Plans and Technological Roadmap

(A) Major Research Contents and Key Scientific Issues

(1) Patterns of land development and spatial governance systems. The key issues include the criteria of "Major Function Oriented Zoning" at different scales, the methods of adjusting the "major function zones" in different development stages and the mechanisms for the coordination and complementation between function oriented zones and policy target areas[176]. To be more specific, the following issues should be addressed:

- The dynamic process of spatial development of land and the spatial-temporal change of regional functions;
- Assessment systems of "major function zones" in terms of finance, investment, land, and population policies and achievement;
- Policies for problematic regions including resource-based cities, undeveloped regions and ecologically degraded areas.

(2) Monitoring and regulating systems for population, resources and environment. The key issues include the evaluation and prediction systems for regional development based on standardized regional units, the key thresholds for judging the degree of regional sustainability, and effective ways for the government to employ controlled resources such as lands, public investment, *etc*. The following issues should be addressed:

- Dynamic monitoring, assessment and policy supporting systems for regional development;
- Security assurance and regulation of resources and environment in the period of high-speed economic growth and urbanization;
- Social and economic problems of population decrease and aging population and countermeasures in the transition period;
- Management of economic development, disaster prevention and environmental preservation of new spatial resources[177].

(3) Distribution and coordination mechanisms of benefits among regions.The key issues include the evaluation on the comprehensive benefits of economic development and environmental preservation, the maximization of the benefits and the distribution of benefits among regions.

- The standard, subjects and methods of the cross-region systems of ecological compensation and transfer payment;
- Watershed management and optimization of water resources allocation;

- Policies for optimizing energy structures under the low-carbon economy and the market mechanism of carbon emission reduction;
- Duties and responsibilities in regional and international cooperation.

(4) Regional administrative system reforms. The key issues include the standard and optimal distribution of public investment and public service facilities, the evaluation on the effectiveness of public services, *etc*. The following issues should be studied in depth:

- Integrated management system of the regional policy target areas and administrative areas;
- Policies for public investment and the standard and optimal allocation of public resources;
- Achievement assessment system of regional management;
- Legal system of regional management and regional legislation.

(B) Roadmap Design Roadmap of Regional Management Systems

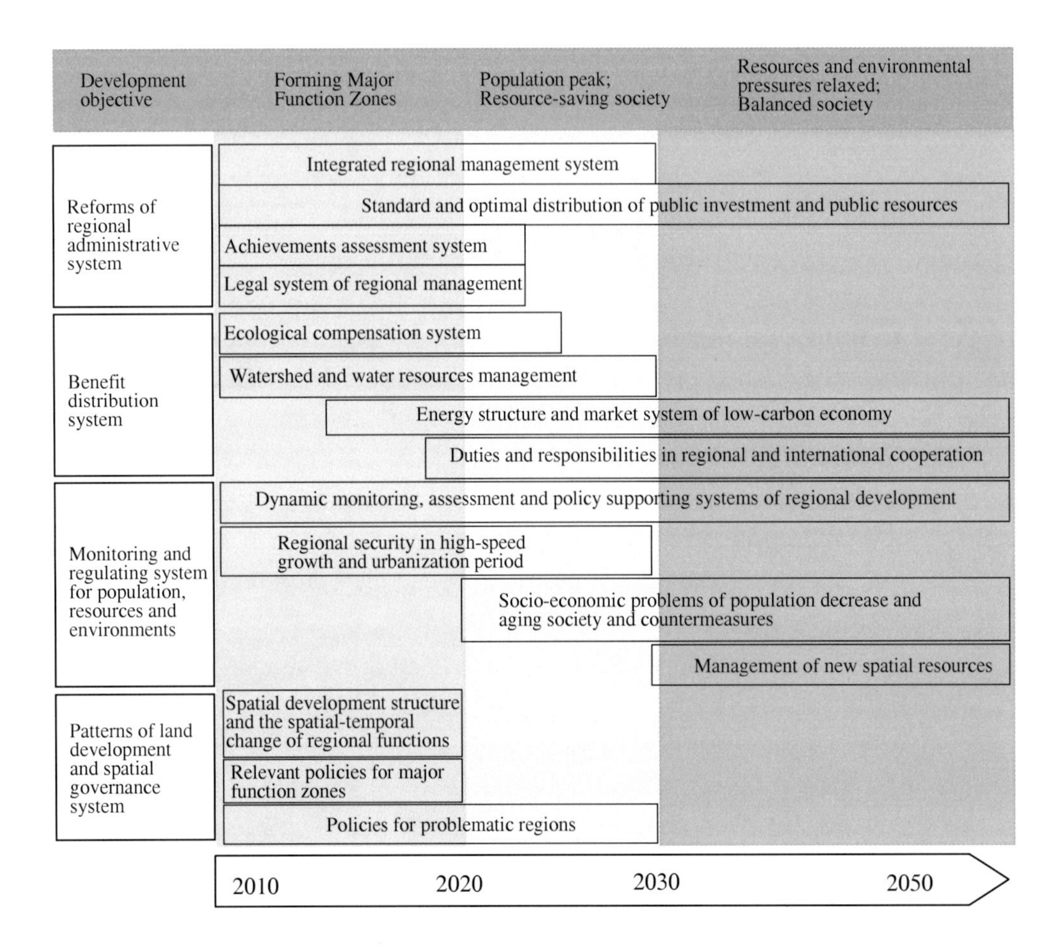

App. Fig. 1.9 Roadmap of regional management system and mechanism research (2010–2050)

Appendix 2

Studies on the Roadmap of Scientific and Technological Development in the Development of Typical Regions

A2.1 Developed Eastern Coastal Regions

A2.1.1 Status Quo and Research Progress in the Development of Developed Eastern Coastal Regions

(A) Status Quo of Development

Eastern coastal regions, as one of the two main axes of productivity distribution in China, enjoy the best development conditions, the strongest economic strength, the most developed industrial basis and the highest intensity and internationalization of cities, yet they also suffer from the most serious man-earth relationship and prominent problems for further development.

As some synthetic studies on the social and economic development of the eastern coastal regions show, the status quo of their comprehensive development has four obvious features: Firstly, the role of eastern coastal regions as gateway cities is very prominent; however, their global/regional functions are not strong. As the forefront in China's opening to the outside world, coastal regions account for over 90% of the total national imports and exports, and more than 80% of the total actual utilization of foreign capital. While there are also problems such as weak global/regional functions, lower level industrial structure, small proportions of high-end products and the service sector, and insufficient ability to distribute global resources. Secondly, although there are three major urban agglomerations, their regional coordination abilities are still relatively weak. More than 70% of the metropolitans of the country are concentrated in these coastal regions where three major urban agglomerations are formed, covering Bohai Bay, the Yangtze River Delta and the Pearl River Delta. However, the three major urban agglomerations still have relatively weak polarization and leading role, strong regional competitions and weak regional cooperation, and poor regional coordination and development capacities. Thirdly, these regions have solid industrial development bases but insufficient innovation capability. Coastal regions are the birthplace of the national modern industries, township enterprises and have a good agricultural foundation, appropriate development level of manufacturing industry, all-round development of the service sector, as well as relatively developed heavy chemical industry, high-tech industry and processing and manufacturing industry. However, compared with the international advanced regions, their overall industrial innovation capabilities are poor. Their development depends mainly on the transferred mature industries from abroad while have a little on research and development (R&D) technology; the labor-intensive industries still occupy a larger proportion, and industrial development is not high level[178]. Fourthly, they have favorable development conditions but poor carrying capacity of resources and environment. These coastal regions have flat terrain, agreeable climate, favorable natural conditions, a large number of sea ports and unique locational conditions for economic development. However, the high-speed industrialization and urbanization in these regions have also resulted in prominent man-earth conflicts as well as a weakening carrying capacity of resources and environment.

(B) Research Progress

The position of the developed eastern coastal regions in the overall China's regional development determines that their regional development issues is always at the core of the researches concerned, which involve economic globalization, cities and urban agglomerations, industrial structure and evolution, regional development mode and spatial evolution, ports and major infrastructure construction, coordination between resources and environment, dynamic simulation and planning, *etc*., and highlight research characteristics of man-earth relationship and laws and mechanisms of regional complexes.

Key research fields include: 1)The impact of globalization and international competitiveness: the impact of economic globalization on the economic, social and industrial structure changes and the spatial agglomeration and diffusion of eastern coastal cities, functional positioning of coastal metropolitans, regional innovation ability and competitiveness, *etc*.[35] 2) The formation, evolution and mechanisms of urban agglomerations: the formation and dynamic process of different types of coastal urban agglomerations, motivation mechanism of spatial agglomeration and diffusion at different levels, and optimization models of space structure, *etc*.[16,179] 3) Industrial structure and spatial evolution laws: the evolution of industrial structure, industrial layout and regional division of labor, changes in and driving mechanism of industrial distribution patterns, *etc*.[180] 4) Regional development models and spatial coordination: regional development models such as those of Wenzhou, southern Jiangsu and Dongguan, causes of and development strategies for the underdeveloped costal regions, and spatial coordination of regional development, *etc*.[38 5) The construction of major infrastructures and port groups: development models, evolutionary mechanism, spatial layout and optimization of international coastal container hub ports, evolutionary mechanisms and models of port system, division and optimization of the functions of port groups, layout of port back-up areas and logistics park, *etc.*[181] 6) Regional development and resource-environment coordination: the process, driving mechanism and trends of land use/cover change (LUCC) in the context of rapid industrialization and urbanization, interaction between human activities, resources and environment, *etc.*[182,183] 7) Dynamic monitoring, simulation and planning: the remote sensing methods for dynamic monitoring of the economic and social development and related resources and environmental factors, simulation and forecast of economic and social development trend through modeling analysis, planning system of different types and levels of rural, urban and regional development, *etc.*; regional planning of Beijing-Tianjin-Hebei metropolitan area, regional planning of the Yangtze River Delta, revitalization planning of Northeast China, and urban agglomeration planning for the Pearl River Delta, *etc.* have been worked out[184].

A2.1.2 Development Trend and Scientific and Technological Demands of Developed Eastern Coastal Regions

(A) Overall Development Trend

The eastern coastal regions will continue to play a leading role in China's economic rise and internationalization. Generally speaking: Firstly, they deepen the degree of internationalization, and will play a leading role in developing into a globalized region in China. Coastal regions are experiencing a large-scale industrialization, urbanization, internationalization, and marketization process with a sustained and rapid economic growth, which has substantially strengthened their development strength and enhanced their geo-economic status in East Asia. With the deepening global economic integration and China's further expanding of opening to the outside, major coastal cities will gradually become internationalized, and take the lead in developing into a globalized region in China. Secondly, the process of urbanization will continue to be accelerated, and the eastern coastal regions will become the most important one for population agglomeration in China. The development of urban agglomerations in the coastal regions in China will still maintain strongly vital in the future. The coastal regions will continue to be a main inflow area of floating population, and become the most economically prosperous regions in China along with the accelerated process of urbanization. Thirdly, man-earth relationship will still be further strained, and resources and environment will suffer much greater pressure. Resources and environmental problems of coastal regions will become increasingly prominent; shortage of resources, excessive consumption of resources and destroy of ecological environment brought by economic growth will make the situation more serious. Fourthly, these regions will exert more impacts on the regional development pattern in China and their leading role will be strengthened. With the enhancing of their comprehensive strength, coastal regions will contribute to strengthen the economic power and comprehensive national power in China, to better support the development of central and western regions, and to promote the development of all regions in China.

(B) Scientific Technological Demands

Coastal regions need strong scientific and technological supports in the process of taking the lead in achieving modernization and internationalization. Firstly, to support coastal regions as the leading one in achieving globalization and to enhance international competitiveness. With their increasingly strong involvement in economic globalization, the economic development of coastal regions will be confronted with increasingly complex and changeable international environment; globalization will exert unpredictable impact on the development of coastal regions in different scales (such as global, national, regional, local scales, *etc*.), and we have to turn to internationalization pattern of regional development and to foresee the changes of role and status of coastal regions in the global economy in the future. Secondly, to guide coastal regions to adjust regional structure and change development mode. Coastal regions, especially the Yangtze River Delta and the Pearl River Delta, have experienced a rapid economic growth over the years. However, with their early-development and policy advantages, the economic development of the coastal regions has been restricted by land, resources, labor costs and national macro-control policies, and they have encountered with development bottlenecks. Thus, it is necessary for those regions to achieve energy saving and emission reduction, to enhance innovative capability, to change the economic growth mode, and to form new impetus of economic growth so as to ensure their core regional competitiveness. Therefore, a range of issues including structure adjustment and change in development modes, *etc*., have become important topics for the economic development in coastal regions. Thirdly, to ensure the coordination between regional development and resource environment and achieve the harmony between man and nature. Along with the industrialization and urbanization process being further accelerated, the population density being further increased, and the industries being further agglomerated in coastal regions, the shortage of development space will be an inevitable trend, and resources and environmental problems will be more prominent. At the same time, coastal regions are sensitive to the impacts of global environmental changes, such as climate warming, sea-level rise, exacerbated coastal disasters, *etc*. Thus, we should make efforts to make those regions adapt to the environmental changes, by carrying out dynamic analysis of the evolution situation of the man-earth relationship in coastal regions, and to solve resource and environmental problems so as to guarantee the harmonious and sustainable development of coastal regions, especially for those with rapid urbanization and industrialization. Fourthly, to promote spatial agglomeration and optimize development pattern. To intensively utilize the limited space for development and to promote the optimization of spatial development pattern are the key to ease man-earth contradictions in coastal regions. The expansion of urban and industrial space occupies a large number of cultivated lands and ecological sites, which threatens food security, and breaks the natural ecological balance. Therefore, it is urgent to dynamically monitor the changes in spatial pattern, to comprehensively consider the development and protection of all types of regions, to divide the regions into different zones in accordance with their spatial development functions, to comprehensively plan and allocate development and protection space, and finally to achieve the intensive spatial development as a whole.

A2.1.3 Overall Goals and Stage Objectives (to 2050)

(A) Overall Goals

In view of the major scientific and technological demands in coastal regions and the international research frontier in this field, we should focus on the natural conditions, material support, socio-economic process and eco-environmental effects of the sustainable development of coastal regions. The researches will concern the driving factors, evolution mechanisms, changes in spatial patterns and development models of the development of coastal regions from three levels, *i.e.*, macro strategy, meso pattern process and micro mechanism so as to reveal the development laws and scientific development models of coastal regions, provide strong scientific and technological support for the sustainable development of coastal regions, and enrich and develop the theories and methodologies of the comprehensive regional researches and regional spatial planning.

(B) Stage Objectives

a. Short-term Objectives (to 2020)

To discover regional responses to globalization, urbanization and industrialization and spatial process, and to establish spatial identification system of territorial functions, to explore regional material and energy flow as well as the processes and mechanisms of agglomeration and diffusion of population and industrial factors, and to form network and platform for data collection and information to support the dynamic monitoring of regional development. In the aspects of the evolution mechanism and analysis and simulation of the regional system of man-earth relationship, we should try to reveal the dynamic mechanism and evolution laws of the regulatory optimization of man-earth system, and preliminarily form a platform for analyzing and simulating regulatory optimization state and scenario of the regional man-earth relationship. In the aspects of the evolution laws and regulation of regional development, we should explore regional development mechanism and laws of spatial pattern change, propose coastal function-oriented zoning schemes and sustainable development models of urban agglomeration, and establish theoretical and methodological system of regional planning and the comprehensive evaluation on major projects. In the aspects of urbanization and the evolution mechanism of land use structure, we will clarify the process of urbanization and the trends of urban and rural structure evolution, and reveal dynamic mechanisms of urbanization and the effects of the spatial expansion of land use on resources and environment, and propose approaches and measures for the overall development of urban and rural areas.

b. Medium-term Objectives (to 2030)

To research the spatial pattern and evolution process of human factors as a clue, scientifically understand the working mechanism of new factors on regional development, and elaborate the effects of the spatial-temporal evolution, mechanism and policies of man-earth relationship on social economy and eco-environment; to build theories of regional man-earth relationship system as well as theoretical system and integrated approach of regional development and urbanization, establish the status of human driving factors in the scientific researches on earth system, and to make them play a role in regional development; to reveal the mechanism of regional sustainable development, evolution laws of development pattern, and approaches of regulation and control, and to set up functional regions of coastal development and the regional planning of these regions as well as the regional policy support system.

c. Long-term Objectives (to 2050)

To continuously track the driving factors, resource-environment foundation, development laws and scientific models of the development of coastal regions from the three levels of macro strategy, meso pattern process and micro mechanism, to explore the influence mechanism of new and old factors on the development pattern of coastal regions and laws of their dynamic evolution, to clarify the scientific mechanism and technical approaches of constructing an economical, harmonious and efficient socio-economic spatial structure, and to provide mature scientific and technological supports for the development of coastal regions.

A2.1.4 Implementation Plans and Technological Roadmap

(A) Key Contents

(1) Regional division of labor and spatial pattern under the background of globalization: the degree of economic globalization established on the basis of marketization is being increasingly deepened. It will exert unpredictable impacts on the development of coastal regions at different levels and lead to the restructuring of international division of labor in different regions. Therefore, it is necessary to dynamically monitor and study the changes in the status, patterns and international relations of costal regions in Asia-Pacific regions and even in a larger regional scope in the context of globalization, and to set up an international development pre-warning system for regional development.

(2) Population flow and evolution of urban spatial pattern in urban agglomerations: in the context of globalization, marketization and informationization, coastal regions will be the first to experience profound

and pioneering changes in the spatial pattern of urbanization. Therefore, it is necessary to carry out dynamic monitoring and in-depth researches on the population growth and population flow in coastal regions, to study regional urbanization, changes in the spatial distribution of urban agglomerations and the corresponding economic and environmental effects, and to direct a reasonable and orderly flow of population.

(3) Industrial structure evolution and locational selection: economic globalization and new technological revolution are driving the restructuring and upgrading of industries all over the world. Therefore, it is necessary to research the evolution of industrial structure in coastal regions, process and laws of spatial changes, as well as the location selecting and optimization mechanism of spatial pattern of industrial structure evolution.

(4) Construction of international shipping center and spatial organization of the functional division among port groups: in a macro-scale, to research the competitions and cooperation among global ports, as well as the evolution laws and mechanism of global shipping centers, and the spatial organization of the functional division among port clusters for the construction of international shipping centers; in a micro-scale, to study the strengthening of the accessibility of port back-up areas and the performance evaluation of port logistics operations.

(5) Rational utilization of coastal land and water resources and coordinated regional development: to explore the spatial-temporal evolution and humanistic influence mechanism of the quantity and quality of land and water resources in coastal regions, to establish a quantitative relationship model for regional development and the evolution of land and water resources, and to simulate and reveal the influences of regional development on the quantity and quality of land and water resources as well as the future trends of land and water resources; to explore regulatory optimization and planning design methods for the coordination between regional development and water and land resources .

(6) Eco-environmental protection in coastal regions and regional development: to evaluate the influences of output pollutants from regional population and industrial agglomerations of different modes, scales and patterns and their biogeochemical process on the changes in ecological, water and macro environment, to explore approaches and measures to control the total amount of pollutants output in the process of population and industrial agglomeration and to improve the quality of water and atmospheric environment, and to establish a quantitative diagnosis, evaluation and pre-warning model for the ecological and environmental security of coastal regions.

(7) Dynamic monitoring and plan designing for the development of coastal regions: carry out mechanism simulation of urban and regional evolution, to analyze problems for decision-making, and spatial policy

simulation, to conduct theoretical inference and trend forecasting, to make real-time analysis on regional environmental and population management, and to achieve the forecast of the evolution of costal development patterns as well as of ecological environment, *etc*.

(B) Research Scheme

Take urban and industrial spatial expansion in coastal regions, as well as the agglomeration effects, ecological impacts and regulatory optimization as a clue, to select typical and important development areas of coastal regions to take the lead in involving into the global economy, to establish a pilot area for the comprehensive research of man-earth system, to carry out long-term follow-up survey, monitor, and trials on various natural and humanistic factors and their interactions and effects from a multi-disciplinary integrated perspective, to establish a data integration and policy -making platform with the functions of data integration, analysis, simulation, policy support, *etc*. for the development of coastal regions, and to break through the bottlenecks of evolution simulation of the complex man-earth relationship system; to further research basic process, and reveal spatial evolution laws, effects and efficiency of factor agglomeration and its impacts on environment and resources, to clarify the spatial optimization model and approach with consideration on both efficiency and fairness, to explore the health indicators and standard system of a harmonious man-earth system, and to strive for achieving a breakthrough in the theories and methodologies of comprehensive researches and spatial planning on regional man-earth system.

Recently, takeing research on spatial optimization and regional coordination mechanism of urban agglomeration as a focus, to analyze main features of the urbanization, spatial distribution laws, key influencing factors and evolution trend of urban agglomerations, to compare effects of different spatial patterns of urban agglomerations on resources and environment, diagnose and evaluate the situation of industrial and population agglomerations and the evolution process of the distribution of lands for construction in urban agglomerations with consideration of the cases of typical urban agglomerations such as the area encircling Bohai Bay, the Yangtze River Delta, the Pearl River Delta, *etc*., to put forward a coordinated development mechanism for urban agglomerations, and provide scientific basis and policy -making references for regional main function orientation and the making of urbanization strategies adaptive to regional resource conditions .

(C) Roadmap (App.Fig.2.1)

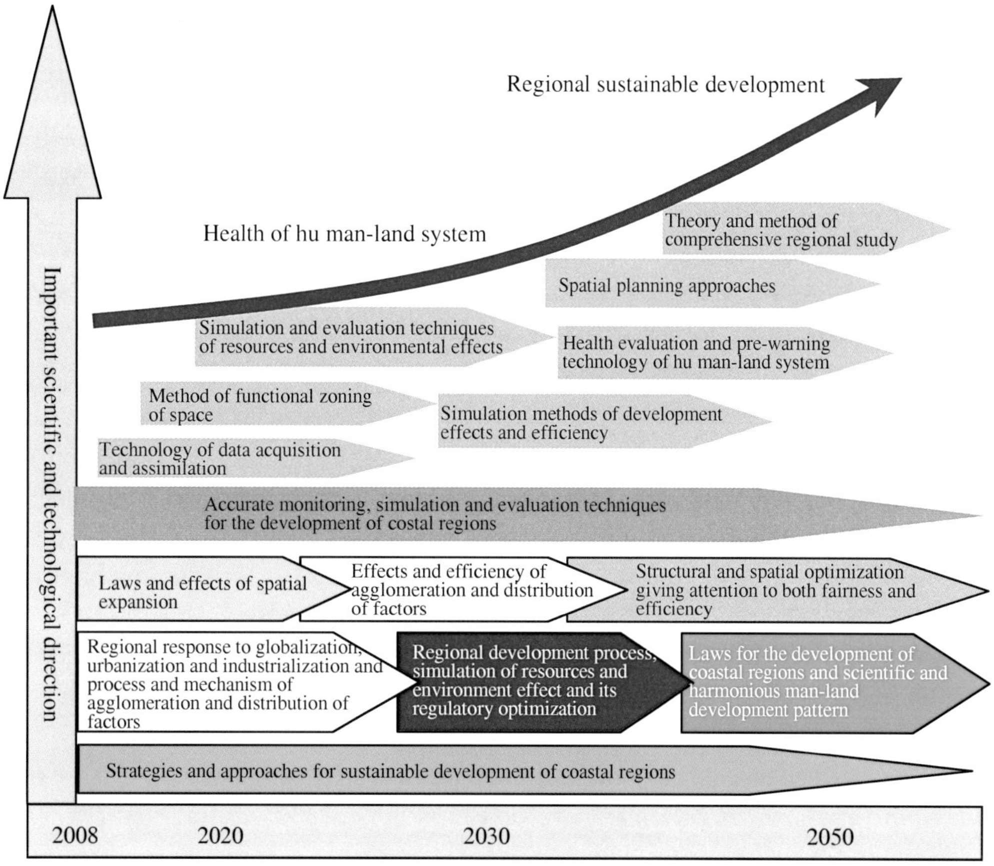

App. Fig. 2.1 Roadmap for research on the development of developed eastern coastal regions to 2050

A2.2 Revitalization and Sustainable Development of the Old Industrial Bases

A2.2.1 The Status Quo Research Progress

In the process of the world industrialization, most of the old industrial bases have experienced a cycle of birth, growth, decline and revitalization. Based on the abundant natural resources, the old industrial bases(OIBs) have made great contributions to regional industrialization and urbanization. However, with the coming of the post-industrialization age, irreconcilable conflicts occur between the traditional development mode of the old industrial bases and the new demands put forward by the economic globalization and informatization. The development of the old industrial bases, as a result, began to fall into a bad situation[185], which has brought about a series of regional economic, social, environmental problems. Therefore, the revitalization of the old industrial bases has become an important issue for researches in the field of regional development all over the world[186,187].

There are two main characteristics of the old industrial bases in China—priority to the development of heavy industry and the highly centralized planned economic system, which accelerated the initial development of the old industrial bases in the early period of the People's Republic of China. However, since the reform and opening up, the inherent defects of the traditional planned system and structural contradictions have restricted the development of the old industrial bases. They have gradually lost their vitality and headed for a recession. From the 1980s, the central government had taken a series of measures for reconstructing and revitalizing these regions, based on the technological upgrading of the state-owned enterprises. As a result, the effects were very limited for it failed to fully realize the fact that the revitalization of the old industrial bases is a systematic, complex and long-term regional development project[188]. In 2003, the Chinese government put forward the strategy of "revitalizing the old industrial bases in Northeast China and other areas", which symbolized that the old industrial bases, as typical "problematic regions", began to be incorporated into the important agendas of the regional development in China, and the implementation of a series of policies has accelerated the revitalization process of the old industrial bases of China[189]. At present, the old industrial bases have made tangible achievements in economic and social development. The economic structure has been improved, the economic growth rate has increased largely, and substantial changes have been made towards market-oriented economy in recent years. The reform and reconstruction of stated-owned enterprises have been in the deepening stage, and the innovative capabilities of the enterprises have been greatly strengthened. Social development has been enhanced in terms of regional infrastructure as well as urban and rural construction, and the regional collaboration mechanism has begun to be established. In spite of the above achievements, the future development of the old industrial bases in China still faces many challenges. For instance, the reconstruction of state-owned enterprises is still to be further deepened, the mechanism of regional integration development has not been established yet, regional revitalization is far from balanced, huge risks still exist in regional eco-environment, there is still a long way to go for the transformation from resource-based cities. Especially, under the influence of the current world financial crisis, the revitalization of the old industrial bases will be challenged by new difficulties.

The studies on the revitalization of the old industrial bases in China has gradually shifted from the initial technology upgrading to the present socio-economic reform, industrial upgrading and structure reconstruction. Their concern has expanded from the economic revitalization of the old industrial bases to fields such as construction on harmonious society, eco-environment management, and industrial histories and cultural in heritage conservation, *etc.*[190] The main research topics include: 1) Reconstruction of old industrial cities, concerning the mechanism and models of the upgrading and reconstruction of old industrial bases in urban areas, such as the case studies of Tiexi model in Shenyang, Yangpu old industrial area in Shanghai, industrial areas along the Yangtze River in Wuhan *etc*. 2) Transformation and sustainable development of resource-based cities, focusing on the changes of relationship between mineral resources and the cities as well as on the suitable economic and urban transformation models for those cities. 3) Shantytown renovation and urban poverty issues, highlighting people as the main body of the urban space, the continuation of community history and social culture, and the residential space, employment, neighborhood, relationship and public space construction of the low-income classes. 4) Protection and utilization of the historical and cultural heritage. 5) Management of eco-environment and construction of the Eco-industrial Park, *etc*.

A2.2.2 Future Development and Scientific and Technological Demand of Old Industrial Bases

The revitalization of the old industrial base in Northeast China, the biggest one in China, has made great progress ever since the revitalization strategy was carried out in 2003. Currently, despite of the world financial crisis and economic depression, there are no fundamental change in the conditions for the revitalization and sustainable development of the old industrial base in Northeast China. Moreover, owing to the solid economic foundation, strong anti-risk capacity and broad leeway for the development of the old industrial base in Northeast China, some of its advantages will even be reinforced in the long run. According to the *Plan of Revitalizing Northeast China* made by the State Council, the overall revitalization of the old industrial base in Northeast China will be generally accomplished in 2020. At that time, it will become a new economic growth pole in China, and some regions will take the lead in realizing modernization. And the role of the old industrial base in Northeast China in building an over-all well-off society as well as in realizing modernization will be more important than ever.

The revitalization of old industrial bases is a comprehensive and complicated regional system project which entails in-depth and systemic researches on a variety of scientific issues so as to realize the revitalization and sustainable development of the old industrial bases.

(1) To scientifically judge the future revitalization conditions and prospect forecast of the development of old industrial bases. The revitalization of old industrial bases is influenced by both domestic and international conditions. Nowadays, the impact of economic globalization and global environmental change on the development of traditional old industrial areas has become a new scientific issue involving the identification of new factors, the processes and interaction mechanisms, and their influence on the development prospect of the old industrial bases. Furthermore, it is important to establish an evaluation indicator system, measurement methods and regulation approaches for the

development conditions, status, and prospect and to monitor the whole process of revitalization and development. It is also necessary to further study the impact of new conditions and factors on the revitalization of old industrial bases by innovating related theories and techniques.

(2) The industrial transformation and new industrialization routes of old industrial bases. After a long period of development, there have generally formed a traditional industrial system mainly based on natural resources exploitation and raw materials in the old industrial bases. However, confronted with the new development trends of globalization and informatization, the revitalization and development of the old industrial bases require a re-consideration on the relationships between these industries and resources, technology, and market. In addition, more efforts should be made to foster new industries, upgrade and transform industrial structure, establish competitive industrial system, and seek new industrialization approach.

(3) To regenerate old industrial cities and implement new urbanization strategy. Old industrial cities are the main entity of old industrial bases, so the revitalization of the former is the essence of revitalizing the latter. In view of this, the leading function of these cities, urban economic zones, and the core cities in the revitalization of old industrial bases should be investigated from the macro-perspective of global city networks. It is also necessary to explore the new urbanization strategy for the revitalization of old industrial bases so as to realize the interactive and coordinated development of new industrialization and new urbanization. Specific comprehensive strategies should be made especially for the reconstruction of the old industrial areas in large cities and the transformation of resource-based cities.

(4) The routes and strategies for old industrial bases to participate in economic globalization. Affected by the Soviet theory of territorial production complex at the initial stage of construction, independent and comprehensive industrial system was the main goal pursued by these old industrial bases, which led to the relatively weak cross-regional economic cooperation under the present market economy conditions. In front of the economic globalization trend, the old industrial bases have few opportunities to take part in, and the industrial and technical connections between them are also relatively limited. In spite of their solid industrialization foundation, the old industrial bases can hardly benefit from economic globalization and even have the risk of being marginalized. Thus the revitalization must be based on the macro-background of economic globalization, and it is important for the bases to search for routes for participating in the global industrial division and to identify regional functional orientation and their role in economic globalization in the new stage[5] .

(5) The resource security and environmental effects of revitalizing old industrial bases. Over the years, the extensive economic growth mode under which economic benefits are achieved at the cost of resource consumption and environmental pollution has brought serious resources and ecological problems to the old industrial areas, and the ecological restoration and environmental

treatment of some regions cannot be accomplished in a short period of time. Meanwhile, the contradiction between resource demand and resource security will be more prominent. Therefore, more attention should be paid to the investigation of proper ways of resource exploitation and utilization and to the employment of both the domestic and foreign resources to solve the long-term resource security problem in revitalizing old industrial bases. The relationship between socio-economic development and eco-environmental protection should be rationally dealt with in making regional development strategies and regional plans, and different roles should be played by different regions. The environmental effects of revitalizing old industrial bases should be evaluated and analyzed with respect of the interactions among economic development, social security and eco-environment. The old industrial bases are also areas sensitive to global environmental changes, so we should make more efforts to comprehensively investigate the interactions between regional human dimensions and natural factors and to explore adaptation strategies and development modes in accordance with global environmental changes[191].

(6) Classified regional guidance policies for revitalizing old industrial bases. It is necessary to analyze and evaluate the effects and problems of the revitalization policies for old industrial bases[192]. We should explore the differential regional policies according to the actual conditions and existing problems of old industrial bases by taking into account the national policy demand and sticking to the principle of differential treatment and classified guidance so as to provide scientific basis for the formulation of classified regional guidance policies by the government.

A2.2.3 Overall Goals and Stage Objectives to 2050

With the support of modern methods and techniques of regional analysis, general researches can be carried out with the following objectives: to establish an integrated platform to monitor, simulate and forecast the dynamic states of the revitalization and sustainable development of the old industrial bases; to explore the coupling mechanisms between and the optimal spatial patterns of new industrialization and urbanization in old industrial bases; to reveal evolutionary mechanisms and regional differentiation of the sustainable development in the "Three-Dimensional Objective Space" of economic growth, social development and eco-environment; to uncover the scientific laws of the cycle of growth-decline-revitalization of the development of the old industrial bases; to put forward new development models and strategic measures adaptive to the sustainable development of old industrial bases so as to provide scientific reference for the sustainable development of other regions in China.

To achieve the above mentioned objectives, detailed research works are planned in three stages.

Major objectives by 2020 include: to build a dynamic monitoring system for the developing state to scientifically evaluate the supporting conditions for the sustainable development of the old industrial bases; to elucidate the coupling mechanisms between new industrialization and urbanization; to optimize territorial functional patterns and urban spatial system; to construct a regional innovation system in old industrial bases; to clarify the coordination mechanism between socio-economic development and resource environment; to explore management models of ecological security and environmental risk; to probe into the evolutionary laws of the decline and revitalization of old industrial bases; and to put forward different strategies and policies for the revitalization and sustainable development of old industrial bases.

Major objectives by 2030 include: to improve the dynamic monitoring system for the sustainable development states and build a platform to simulate and forecast the development states of old industrial bases; to analyses the spatial-temporal processes and dynamic mechanism of regional transformation; to explore the social, economic and environmental effects of revitalization policies and discover the evolutionary mechanisms and regional differentiation in the "Three-Dimensional Objective Space"; to clarify the micro-mechanism during the process of growth-recession-revitalization of old industrial bases; to scientifically understand the effect mechanism of new factors; to reveal regional response mechanism of old industrial bases to economic globalization and global environmental change; to detect the territorial function transformation of old industrial bases under new conditions; and to put forward new modes and strategic approaches for the development of old industrial bases.

Major objectives by 2050 include: to further improve the integrated scientific platform of the dynamic monitoring, simulating and forecasting for the sustainable development; to continuously track and systemically study the new driving factors, resources and environmental foundation as well as the evolutionary law and scientific development models of old industrial bases; to construct regional risk identification system; to explore the scientific mechanisms and technical measures for the functional optimization of the spatial structure of old industrial bases; to identify major technical and economic parameters and their dynamic changes; to put forward new development models for the old industrial bases adaptive to the new environment so as to provide scientific reference for the sustainable development of other regions in China.

A2.2.4 Implementation Plans and Technological Roadmap

(A) Key Scientific Issues

With a view to national and regional demands, as well as the overall development goal to 2050, the scientific issues regarding the revitalization and sustainable development of old industrial bases are as follows:

(1) Assessment and forecast of the revitalization situation;

(2) Regeneration of old industrial cities and spatial pattern of new urbanization;

(3) Industrial transformation and new industrialization routes;

(4) Construction and optimization of national grain security base;

(5) Resource-environment carrying capacity and ecological security management;

(6) Regional innovation system based on the application of new techniques ;

(7) Sustainable development modes under the background of globalization and global changes.

(B) Research Contents

According to the scientific issues mentioned above, the main research contents can be summarized as follows. Firstly, to diagnose and evaluate the revitalization situation and to forecast the prospect of the sustainable development of old industrial bases. Secondly, to develop the spatial pattern and routes of new industrialization and new urbanization of old industrial bases.Thirdly, to identify the carrying capacity of resources and environment and make rational spatial governance on the revitalization of old industrial bases. Fourthly, to explore new modes and strategies under new conditions and the effects of new factors. Fifthly, to probe into the laws of the economic transformation and regional disparity of old industrial bases and reveal the pattern and evolutionary laws of regional functions.

(C) Research Sheme

On the basis of the key scientific issues on the revitalization and sustainable development of old industrial bases, three scientific programs focusing on the revitalization of the old industrial bases in Northeast China are to be carried out in order to meet the great scientific requirements of the government.

(1) To research the mechanisms of old industrial base revitalization and innovative development. According to the functional orientation and overall revitalization goal of the old industrial bases in Northeast China, the aims of this program is to diagnose and evaluate the revitalization and sustainable development situation, to reveal the spatial-temporal rules of the revitalization and economic transformation, and to put forward the sustainable development modes of old industrial bases. The focuses of this program are on the diagnosis and assessment system of the revitalization situation of the old industrial bases in Northeast China, the coordinated development mechanism between socio-economic development and eco-environment protection, the strategies and modes of new industrialization and new urbanization, the regional innovative system and the ways of economic integration, the regional environmental risks and the ecological security pattern, *etc*., among which the key scientific issues are the sustainable development and innovative modes of the old industrial bases. The main innovative point is the exploration of the coordinated mechanism between the revitalization of the old industrial bases and the

resource-environment system.

(2) To find new urbanization mechanism and spatial pattern of the old industrial bases in Northeast China. In view of the complex mechanism problems in the urbanization process under the background of the overall revitalization of old industrial bases, this program is to scientifically identify the key impetus of urbanization and their working mechanisms, to forecast the future prospect of urbanization, to analyze the spatial agglomeration and diffusion pattern of urban agglomerations, and to reveal the evolutionary laws of the spatial structure of regional urbanization. The objective of this program is to put forward a new urbanization pattern and corresponding optimization schemes for the old industrial bases in Northeast China. The focuses of this program include the process, dynamic mechanism and trend of the urbanization in the old industrial bases in Northeast China, the reconstruction and transformation of resource-based cities of the old industrial areas in large cities, and the strategies and approaches of the coordinated development of urban and rural areas, *etc*., among which the key scientific issue is to identify the main impetus for the new urbanization of the old industrial bases in Northeast China and their transformation mechanism. The innovative point is to put forward the spatial optimization scheme for the new urbanization of the old industrial bases in Northeast China.

(3) To construct and optimize the grain security bases in Northeast China. Based on the new national and international conditions of food security, the main goal of this program is to establish national grain security bases and strengthen the supporting capacity of Northeast China to the grain security in China. It analyzes the main problems and restriction factors in the construction of the grain security bases in Northeast China and its focuses are on the investigation of the comprehensive potential of grain production of Northeast China and its influencing factors, the forecast of the supporting capacity of Northeast China to national grain security, the coordination of the relationships between the development of the processing industry of agricultural products and national grain security, the discussion on the spatial restructuring plans and development modes of grain security bases in Northeast China, and suggestions on the strategies and countermeasures of the sustainable development of the grain security bases in Northeast China. The key scientific issue is to forecast the comprehensive potential of grain production and the supporting capacity of Northeast China to grain security. The innovative point is to put forward an optimization and spatial restructuring plan for the construction of the grain

security bases in Northeast China.

(D) Roadmap (App. Fig. 2.2)

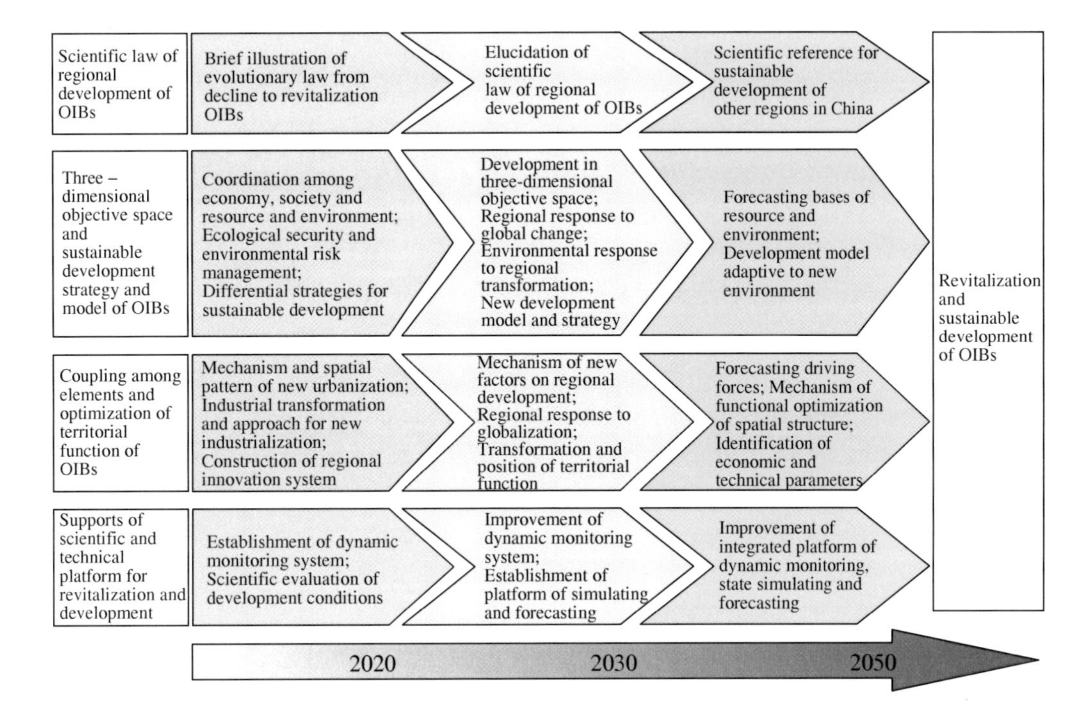

App. Fig. 2.2 Roadmap for the research on revitalization and sustainable development of old industrial bases to 2050

A2.3 Sustainable Development Roadmap for Mountainous Areas of China

China ranks first in terms of area of mountainous regions (plateaus, high mountains and hills) in the world, which accounts for about 70% of its total land area. The remarkable ecological service functions and huge resource and environment carrying capacity of the mountainous areas represent the strategic security nature of the supporting ability of national development. To pay much attention to the development of mountainous areas which supports the future development of China should be regarded as an important component of national strategic management[193].

A2.3.1 Status Quo and Research Progress of Mountainous Areas

The mountainous areas in China, where ethnic minorities are concentrated, are characterized by sharp regional differences, strong complexity, lagging infrastructure, underdeveloped social economy and intense man-earth conflicts due to restrictions of natural conditions and national strength.

Confronted with the realistic demands for the development of mountainous areas, researches on mountainous areas emerged, but the early ones were monotonous in contents and limited in research fields, which satisfied the production demands of national economic construction and the basic living demands of citizens at that time yet seems too narrow-viewed and primary now. Since the beginning of the 1990s, affected by the concept of sustainable development, the researches on mountainous areas have gradually evolved from monotonous and partial ones to comprehensive and holistic ones. However, those researches, as a whole, are still far from satisfactory and systematic. Development strategies are seldom concerned and researches at the national level are also scarce[194].

There have been long intensive man-earth conflicts in the mountainous areas in China, and the sustainable development is confronted with significant strategic issues and challenges. From the perspective of research progress, relatively systematic researches on the population of mountainous areas appeared in the middle and late 1990s, including those on the geographical distribution of population, culture and health, population and ethnonymics, as well as countermeasures on population issues, among which the *Report of Development in Mountainous Areas of China* (2003, 2007) is the most representative one with an in-depth analysis on the mountainous areas and settlements [193,195]. The researches on resources in the mountainous areas mainly concern the biodiversity conservation, rational utilization of water energy resources, development of mineral resources and ecological management, regulation and efficient utilization of water and soil resources, assessment of mountain landscape resources, evaluation on the productivity of forest and grassland resources, etc. Furthermore, natural forest protection, water and soil conservation, and ecological and environmental effects of hydropower projects on rivers have become the hot research topics, and researches on ecological shelter effect of the mountainous areas are drawing more and more attention, among which the researches on ecological shelter of Tibet has made leading progress, with their findings and significant scientific suggestions approved by the State. The economic and social development in mountainous areas has been long positioned as support to national economic construction with guaranteeing the resource demands of the development in plain areas as its main line of development. Researches on the development of mountainous areas have revealed the stages, resource dependency, primary characteristics of industries, and economic development level and defects of the industrialization in mountainous areas in China. Compared with the developed countries, the strategic planning and policy researches on mountainous areas are far from being integrative, coordinative and comprehensive. Especially, the issues such as economic and ecological compensation, comprehensive development of mountainous areas, *etc*. have not been completely solved on the legislative level, which obviously restricts the development of the mountainous areas [193,195].

A2.3.2 Judging the Sustainable Development and Scientific and Technological Demands in Mountainous Areas

(A) Judging Development Trends

Under the influences of globalization and the political, economic and technological factors, the sustainable development of mountainous areas will mainly show the following trends:

(1) Population and urbanization in mountainous areas. In a fairly long period of time, the scale and intensity of the emigration from the mountainous areas to cities and towns should be enlarged and strengthened in order to keep in step with the progress of building a well-off society in an all-around way. On the other hand, along with the population transferring from mountainous areas to plains or to cities and towns within the mountainous areas, migration of population for ecological projects, and outflow of population from impoverished mountainous areas, negative growth of population occurs in some counties in the mountainous areas.

(2) Development and utilization of resources in mountainous areas: The fundamental realities of large population and huge economic gross in China determine its dependency on domestic supply of the major resources for economic and social development. At present, the disorderly development of resources and the lag of researches have restricted the sustainable utilization of resources in mountainous areas[196].

(3) Ecological functions and ecological security in mountainous areas. Due to the sensitivity, fragility and vulnerability of the ecosystem in mountainous areas, to protect the ecological functions of mountainous areas is of considerable necessity and urgency to the guarantee of ecological security in China. The basic and diffusion effects of the ecological benefits in mountainous areas have consolidated the strategic position of the ecosystems in mountainous areas. Moreover, under the influence of globalization, the ecological security issues in the mountainous areas of China will draw global attention gradually due to the wide regional coverage, uniqueness and complexity of these areas.

(4) Economic and social development in mountainous areas. The mountainous areas in China will become the supply base of energy resources to the whole country, especially the supply of hydropower and coal. They will also be established as the production center of mining industry, for example, nonferrous metals, ferrous metals *etc*. Besides, processing industries of building materials, forest products, biological products, traditional Chinese medicine materials, animal husbandry products, *etc*. will hold a strategic position as national bases in China. The mountainous areas will develop into important production bases of organic agriculture and green agriculture as well as major bases of ecological tourism, leisure tourism and recuperation attracting visitors from all over the country and the world. And the social security system and public service system will also be established in these areas[197,198].

(B) Scientific and Technological Demands

(1) Succession law of spatial-temporal patterns of the population and settlements in mountainous areas. On the basis of the carrying capacity of resources and environment, the overall coordinated development of the population and spatial layout in mountainous areas is the foundation on which a well-off society can be built and the sustainable development of mountainous areas can be guaranteed. Since settlement is the most basic unit of both social economy and social organization in mountainous areas, to find out the spatial-temporal evolution law of it is an important scientific proposition that must be highly concerned in the future[195,197,198].

(2) Researches on sustainable utilization of natural resources and environmental health in mountainous areas. The urbanization and industrialization in mountainous areas will definitely cause many conflicts in the utilization of natural resources. The following are the important scientific issues that must be addressed in the development of mountainous areas: construction of basic information platform for resource utilization, regional planning of resources, benefits of the inter-regional flow of resources, responses of resources to global changes, carrying capacity of resources, and management of the sustainable utilization of resources, *etc.*[196]

(3) Researches on the urbanization and ecological security process in mountainous areas. Although the urbanization of the mountainous areas is a process of imbalanced development, it is an inexorable trend that the mountainous areas will experience a rapid, coordinated and overall development. During this process, the following important scientific issues are to be urgently solved: urban expansion and ecological gains and losses in the mountainous areas, comprehensive carrying capacity, grade and scale of urban system, assessment on disaster risks, and the emergency system, *etc*.

(4) Researches on the construction of a low-carbon industrial system in mountainous areas and countermeasures of adaptation to climatic changes. As an organic entity for consumption of resources as well as the generation and reduction of wastes and greenhouse gases, the industries in mountainous areas are considered as an important component of realizing the goal of adapting to global climate changes. Therefore, to promote the eco-transformation of the industrial system and reveal the ecological adaptation laws of it are objective requirements of the sustainable development of industries in mountainous areas.

(5) Researches on eco-environmental influence process in mountainous areas and security guarantee technology. As the mountainous areas act as long-term resources output areas, the following scientific issues are to be addressed urgently: measurement on the sustainability of the ecosystem in the mountainous areas in China, ecological restoration in mining areas, maintenance of ecosystem functions, *etc*.

(6) Researches on the function-oriented zoning of the comprehensive development in mountainous areas. The comprehensive development of

mountainous areas must be based on the concept of sustainable development and the uniqueness and diversity of these areas. Priority has to be given to researches particularly in the following aspects: Firstly, the function-oriented zoning of the comprehensive development in mountainous areas. Secondly, the priority development and promoting mechanisms. Thirdly, the multi-objective standard system of the construction of the well-off society in mountainous areas and the assessment methods. Fourthly, the criteria for Modernized Mountainous Areas in China and future scenarios.

(7) Researches on regional differentiation and coordination mechanism of man-earth relationship in mountainous areas. Due to the vertical differentiation, the man-earth relationship in mountainous areas is quite diversified and complex, which puts forward higher requirements on the coordination work. In view of this, the following important scientific issues need to be solved urgently: behavior mechanisms of varied regional scales for man-earth relationship, productivity changes and evolution of man-earthrelationship, harmonious man-earth relationship and accommodation of ethnic and cultural relations in mountainous areas.

(8) Strategic planning of the development in mountainous areas and the building of social supporting capacity. The lagging development in mountainous areas urgently requires the central government to pay close attention to the important social issues from the strategic height of promoting comprehensive, coordinated and sustainable development of mountainous areas. The government should also adopt positive development policies and strategies, incorporate the development of mountainous areas into the overall planning of national development, and promote all-around and high-level development of mountainous areas through specific strategic planning and regulations[197,198].

A2.3.3 Overall Goals and Stage Objectives to 2050

(A) Overall Goals

The overall objectives are as follows: 1) to carry out fundamental and specific plans of considerable scientific significances as well as major scientific and technological supporting work in mountainous areas in a systematic manner; 2) to promote the construction of the revitalization law in mountainous areas; 3) to complete the strategic planning of the development in mountainous areas; 4) to establish and further improve the construction system and the layout of the productivity in mountainous areas; 5) to make sure that the ranks of cities and towns are compatible with the scales of settlements; 6) to strengthen the capacity of the environment to support development comprehensively; 7) to form the strategic pattern of construction on ecological security; 8) to set up a coordination mechanism of man-earth relationship so as to pave the way for the sustainable development of mountainous areas.

(B) Stage Objectives

By 2020, the level of integrated researches on the development of mountainous areas in China will have been considerably enhanced. And China's leading position in the world in such fields as the settlements in the mountainous areas, division of regional types, the industrialization pattern, ecological industry construction and environmental effects, and urbanization and ecological security, *etc*. should be maintained. By that time, a thorough understanding and control on the natural resource conditions in mountainous areas will have been acquired and a basic information platform will have been established; strategic blueprints for the coordinated development of mountainous areas will be developed; the spatial-temporal evolution laws of population transfer, settlement patterns and urbanization in mountainous areas will be revealed; ecological construction level of industries will have been enhanced and the environmental effect scenarios of ecological industries in mountainous areas will have been simulated at the same time; theoretical and practical research level on the strategies of differences in development as well as on the adaptive modes of development in mountainous areas will be enhanced .

By 2030, in order to meet the requirements of the overall regional strategic pattern of national economic development, a framework for mountain science which will equip China with all the necessary conditions for the scientific researches on the large-scale, systematic and regionalized development of mountainous area will have be built up. By that time, brand new 21st century mountainous areas will be formed step by step in which advanced science and technology, well-educated farmers and new rural organization modes will enable the mountain areas to participate in market competition; new development modes for modern mountainous areas featured by the support from plain areas for the development of mountainous areas and balanced development will be achieved gradually; the productivity layout of modern mountainous areas will be set up; the ecological industry system and the supporting system of policies and institutions on sustainable development which are well adapted to the economic globalization, ecologicalization and international market competition will be established; the theoretical and technological systems of mountain science will have been basically formed[194].

By 2050, the researches on the development of mountainous areas in China will have formed a systematic scientific and theoretical system with focus on the researches on the economic and social sustainability of mountainous areas, ecological security and service functions as well as the earth surface process and the succession theory of the behavior mechanism on man-earth relationship. We should also make efforts to hold the leading position in the world in such aspects as methods for mountainous population prediction, quantitative and qualitative changes of resources and simulation techniques for climatic response, simulation techniques for the spatial-temporal patterns of settlement succession, identification techniques for the urbanization process and ecological security in mountainous areas, ecological industry integration and scenario modes of low-carbon economy, division of regional types and adaptive strategic preplans, *etc*.

A2.3.4 Implementation Plans and Technological Roadmap

(A) Research Tasks during the 12th Five-Year Plan Period (2011–2015)

To construct a basic information platform for the mountainous areas in China: To define the scope and criteria of mountainous areas scientifically, to build up a database of natural and socio-economic attributes with the county administrative division as the basic unit, and to construct a basic information platform for the researches on and management of the resources, economy and society of the mountainous areas of China.

To make strategic plans for the development of mountainous areas of China: Make a general plan by placing the development of mountainous areas in the overall national strategy, to give prominence to the strategic position, division of labors, and objectives of the development in mountainous areas, and to form an eco-compensation mechanism for mountainous areas.

To systematically research the industrialization scenario and ecological industry construction in mountainous areas: To evaluate the agricultural industrialization potential and zoning, and explore the direction and approaches of the agricultural industrialization in the mountainous areas in China. To incorporate the development of mountainous areas into the national industrialization process without requiring the establishment of a complete industrial system there. To carry out researches on the spatial organization and environmental effects of the ecological industries in mountainous areas in time.

(B) Research Tasks during the 13th Five-Year Plan Period (2016–2020)

Researches on the population transfer as well as the succession and reconstruction of settlements in the mountainous areas in China: To figure out the directions, paths and tendencies of the population flow in mountainous areas. To carry out the reconstruction of settlements based on the revelation of the evolution relations and future evolution patterns of settlements in mountainous areas[197,198].

Researches on urban system construction as well as disaster prevention and mitigation system and emergency response capability in mountainous areas: To scientifically understand the characteristics of the rank scale of towns and the disaster risks in mountainous areas, and to establish corresponding urban infrastructure and emergency engineering systems as well as disaster prevention and mitigation systems.

To systematically research the construction of new socialist mountainous areas: To establish evaluation criteria for the new socialist mountainous areas, to define the objectives of the construction of new socialistic mountainous areas, and to set up the law and regulation system for the construction of new socialistic mountainous areas[197,198].

(C) Implementation Plans and Technological Roadmap of the 12th Five-Year Plan

a. Research on Strategic Plan for Development of Mountainous Areas in China

Research objective: To solve the problems of the industrialization, urbanization, and realization of an overall well-off society, settlement reconstruction and ecological compensation in mountainous areas.

Research content: Spatial pattern, urbanization and settlement, industrial potential evaluation, and technical systems of urban disaster prevention and mitigation in mountainous areas.

Technical keys: Correlation mechanism of the carrying capacity of the resources, environment and population in mountainous areas; evaluation on the strategies' supportability for mountainous area development[197,198].

Innovation points: To enhance the capability of analyzing the complex systems in mountainous areas; to grasp the interactive law between the strategic plans and policies for mountainous areas.

Technical route: To focus on the main problems and specify the general approach, objectives and tasks of the development in mountainous areas, to effectively coordinate the overall national planning, specific planning and local planning, and to promote a rapid and comprehensive development of the mountainous areas in China.

Organization plan: To strengthen the scientific researches and implement specific plans of considerable scientific and technological significance; to make implementation schemes for specific scientific and technological plans, project management methods and project application guidelines; to encourage the close cooperation between research institutes in order to form an integrative research force.

b. Construction of Integrated Information System in Mountainous Areas and Fundamental Work

Research objective: To establish an integrated information system in the mountainous areas to provide basic data support for both the national and specific researches.

Research content: Digital basic database and atlas of the resources, environment, disasters and social economy of China's mountainous areas.

Technical keys: Multi-modular support integration system of 3S technology; multi-source spatial information extraction, standardization and spatial-temporal characteristics of mountainous area data; spatial visualization and VR technical integration[199].

Innovation points: To build up the standardized mountainous area database and digital mountainous areas so as to realize the comprehensive digitalization of the mountainous areas; to search for and carry out inversion of integrated mountainous area information under the support of 3S technological multiple modules.

Technical route: To research the mining of multi-attribute spatial-temporal data, information extraction, data storage and analytical methods for the integrated information of the mountainous areas; to establish a comprehensive technical system of digital mountainous areas, and to set up a digital technical simulation platform for the integration of visualization and VR simulation [199].

Organization plan: The Bureau of Science and Technology for Resources and Environment as well as the Scientific Information Center for Resources and Environment, CAS lead in organizing a comprehensive expert team to establish a basic information technology platform in mountainous areas through implementation stage by stage, construction region by region, and system integration.

c. Research on Mountainous area Urbanization, Industrial Agglomeration and Coordination between Evolution and Optimization of Social Structure

Research objective: To construct a structural model for the comprehensive carrying capacity of towns, to build up urban disaster prevention and mitigation system, to select typical low-carbon industries in mountainous areas as a strategy of adapting to the climatic change, and to establish scenario methods and index system for the urbanization and industrialization in the mountainous areas in China.

Research content: Model for the comprehensive carrying capacity of towns in mountainous areas; ranking scale and disaster prevention and mitigation system; mountainous area urbanization and industrialization based on the three-dimensional objectives (economic growth, social security and ecological environment) as well as the linkage mechanism and pattern of social structures.

Technical keys: To set up evaluation methods for the comprehensive carrying capacity of towns in mountainous areas; to reveal the scenario analysis framework of the urban construction and industrial layout under the three-dimensional objectives; to establish ranks of towns and scales of industries, to set threshold values of disaster risk, and to build up pre-warning and forecasting system in mountainous areas.

Innovation points: Based on the existing regional methods, to highlight mountainous area complex, scenario analysis and simulation methods of mountainous area industrialization and urbanization, and the regional development mechanism and trends based on the coupling of the three-dimensional objectives.

Technical route: To forecast the scenarios of industries and urban development in mountainous areas; to carry out researches on the industrial agglomeration, rank scales of towns and the adaptation laws of resources and environment, to probe into the analysis methods for the threshold values of disaster risk and carrying capacity, and to explore the future modes of the mountainous area complex in China.

Organization plan: To further expand the scale, scope and implementation fields of the systematic and integrated researches on mountainous area development; to promote the enhancement of the research level on the industrial and urban development in mountainous areas through significant scientific and technological projects.

A2.4 Scientific and Technological Roadmap for Research on Regional Development in Border Areas of China to 2050

Contrary to coastal areas, border areas refer to the certain marginal areas of the land territory of a country. From both the geographical and the economic points of view, there are five provinces and four autonomous regions which can be defined as border areas in China, *i.e.* the Liaoning, Jilin, Heilongjiang, Gansu and Yunnan provinces and Xinjiang, Tibet, Guangxi and Nei Mongol Autonomous Regions. The extending parts of the border areas and the northwestern minority nationality regions also involve Qinghai, Shaanxi, Guizhou and Sichuan provinces, Ningxia Hui Autonomous Region and Chongqing City. The border areas in China cover 7.6396×10^6 km^2 which accounts for 76.77% of the total land area, with a population of 3.01×10^8 which occupies 22.92 % of the total population in China.

The border areas are the strategic barriers of China's frontier and the pivotal regions for safeguarding the territorial integrity, national security, political security and diplomatic relations with the neighboring countries; they are the vital strategic bases of energy and natural resources and substitute areas of resources in supporting the modernization construction in China and are thus of considerable significance to the security of energy and natural resources in China; they are also the important gateways opening up to the outside world and affect the overall pattern of the opening up of China.

The border areas are the pivots of economic growth in western China, bases of characteristic agriculture, animal husbandry and fruit industry, and important national and international tourist destinations; they also play important roles in realizing the balanced regional economic development in the undeveloped regions and narrowing the economic gap between eastern and western China.

Most impoverished population in China agglomerate in these areas, which hampers the process of realizing an affluent society in China. The ecological environment in the windward regions and upper reaches affects the environment quality in the leeward regions and lower reaches, so the development trend of these border areas affects the overall quality of ecological environment and the process of constructing an environment-friendly society in China. As the development in the border areas is essential to the overall situation in China in aspects such as the energy security, natural resources security, ecological security, national security and economic development, these areas hold an especially important strategic position in all the work of the Party and the State.

A2.4.1 Status Quo and Research Progress in Border Areas

(A) Status Quo of Development

(1) A basic pattern of opening to the outside world has formed in the border areas, covering northwestern, northeastern and southwestern open zones in China. The formation of the development pattern, characterized by taking the cities and towns at various levels as the centers, the border economic cooperation zones, export processing zones, mutual trade regions and various industrial parks as the bases, and the international regional transportation corridors as the connections, has been accelerated, thanks to the "Development of the Western Regions" in China, the construction of "Shanghai Cooperation Organization", China-ASEAN Free-trade Area, and Regional Cooperation in Northeast Asia *etc*.

(2) At present, the largest petroleum and natural gas exploitation and processing base in China, large-scale coal bases, nonferrous metal exploitation and preliminary processing bases, and large-scale hydropower stations are under construction. The exploitation of natural resources has become the main driving force of economic growth in these underdeveloped resource-rich regions.

(3) The border areas lie at the bottom of the economic system in China, where the poverty pressure is huge. Border areas are mostly characterized by poor natural conditions, vast territories with a sparse population or huge mountains, high traffic cost, lagging infrastructures, and backward economy. These areas are also the major regions where the impoverished population in China is concentrated with a high occurring frequency of poverty and harsh natural environment, thus the regional development is seriously restricted.

(4) These areas are windward and upper reach regions with fragile and sensitive ecology. They are confronted with serious environmental problems such as the ecological destruction, water loss and soil erosion, security threat of water source regions, and disappearance of ecological barriers in the sandstorm source regions, *etc*., and the low level of regional economic development has resulted in the more and more prominent conflicts between economic development and ecosystem conservation. Due to the absence of the national counting system of green GDP, the ecological compensation and transferring payment mechanisms between the upper and lower reaches and among different regions have not been developed yet in China.

(5) Although some innovative enterprises have been established in the border areas, the development of industrial clusters is lagged. Moreover, the academic institutions and universities have not been closely connected with local demands, the level of production-education-research integration is low, the regional innovative system has not been formed yet, and its driving effects on regional development is by no means strong.

(6) Peace and development are the mainstreams, but the threat of national splittism from Tibetan Separatist force and Eastern Turkestan group still exists.

(B) Research Progress

The researches on regional development in the border areas mainly concentrate on the following issues: the security situation in the border areas[21-25], the poverty of ethnic economy in northwest and Southwest China[200-204], the opening system in the frontier minority regions, the construction of energy corridors in Western China (Xinjiang) and the energy security strategy of China, and financial investment strategy in northwest and Southwest China. A series of researches on the regional development in Xinjiang have been carried out in the Xinjiang Institute of Ecology and Geography since 1950, including:

(1) Regional development, concerning the regional planning and specific planning in key regions, such as the Regional Development and Renovation of Ürümqi, Regional Development Planning for regions along the North Xinjiang Railway, and Balanced Regional Development in the Economic Zone on the North Piedmont of Tianshan Mountains; Strategic Development Planning for regions along the South Xinjiang Railway and the Petrochemical Industrial Belt; the Regulation Strategy of Administrative Divisions in Xinjiang,"Major Function Oriented Zoning" in Xinjiang; the Balanced Regional Development in the Middle and Lower Reaches of the Tarim River; the Strategies of Balanced Regional Development in the North Tarim Basin; the Macro-strategies of Environment in Xinjiang, and "Study on the Effects of Urbanization on LUCC in Oases", an important orientation project of Chinese Academy of Sciences.

(2)Urban systems, including the researches and planning of urban systems on different spatial scales, such as the Planning of Urban Systems in Xinjiang (2002–2020), the Development Strategy of Small Towns in Xinjiang, the Development Strategies of Ürümqi Metropolitan Circle, the Agglomeration, Diffustion and Regulation of Cities and Towns on the North Piedmont of Tianshan Mountains, Strategic Development Planning of Cities and Towns in the border areas of Xinjiang, the Development Approaches of Urban-rural Integration in Xinjiang, and Strategic Regional Development Planning of Kuytun-Dushanzi-Ürümqi Regions.

(3) Tourism development and ecological conservation, including the studies on the theories of tourism development and ecological conservation, such as multi-level (province, region, county, scenic spot, tourist route) and multi-scale tourism planning like the General Planning of Tourism in Xinjiang (2002–2020), Supporting Theories and Empirical Studies on the Tourism Planning in Xinjiang, the Security Systems of the Natural Heritage Spots, and Development and Protection of the Historic Sites of Minority Nationalities and along the Silk Road.

(4) Central Asia Regional Cooperation, including the study on the Participation of Xinjiang in the Subregional Economic Collectivization of Surrounding Countries, Regional Economic Cooperation and Resource and Ecological Security in Central Asia which further involves the Modes and Countermeasures of Water Allocation of International Rivers, Digital Reconstruction of Eco-environmental Evolution, the Strategies of Regional Economic Cooperation with Central Asian Countries, *etc*.

A2.4.2 Judging the Future Regional Development in Border Areas and the Scientific and Technological Demands

(A) Judgment on Development Trend

(1) The development mode of border areas will gradually shift from advantageous resource export and primary processing bases to the characteristic industrial clusters of deep processing of resources which will become an important pillar industry in supporting the economic development in China, pushing the local economic growth, and helping the local minority nationalities to shake off poverty and become well-off.

(2) The Central Asia, China-Russia, South Asia, Southeast Asia and Northeast Asia international corridors will be built in the border areas to form the diversified and multi-corridor extroverted comprehensive transportation network systems, regional international logistic centers, important merchandise distribution centers, and some important trading ports; some export processing industries will be shifted from the eastern coastal areas to the border areas for the formation of comprehensive export processing bases.

(3) As the urbanization process intensifies the shortage of cultivated land in China, western China will become a key region in a new round of water and land exploration. Projects of inter-basin water diversion, water-soil improvement in the Yellow River Basin, and exploration of water and land resources in northwest and northeast China will be implemented in western China.

(4) The economic development in the minority regions and the ecological environment will be significantly improved, and the "Three Big Gaps" will be gradually narrowed. Various national measures will be implemented to strengthen the self-development capability of the poverty-stricken regions, promote the regional economic development, improve the living condition of urban and rural residents, and gradually narrow the gaps between eastern and western China, urban and rural areas, and the rich and the poor.

(5) Common development objectives in terms of energy resources, supply-demand relationships of the market, and benefit-sharing will be sought by the border areas and neighboring countries to form political, economic security systems and to gradually achieve regional economic integration.

(B) Analysis of the Scientific and Technological Demands

(1) The guarantee system of natural resources in national modernization construction, including how to implement water and land exploitation and water diversion schemes in border areas and how to map out macro-strategies of international relations with the countries in the lower reaches, *etc*. on the basis of the security of cultivated land resources in China and the settlement of ecological emigrants from the headwaters, sandstorm source regions, forbidden areas for development like serious damaged zones, and poverty-stricken regions.

(2) The system for border areas to open to the outside world under the background of globalization: in view of the existing cooperation barriers such

as differences in nation-state systems, degree of opening to the outside world, corridor construction, transportation connection, low efficiency of customs, and trade barriers, *etc.*, to do comprehensively research on how to construct the opening system in border areas from the perspectives of the exploitation of natural resources, market exploitation, the feasibility of cost benefit, national security, and benefit balancing among countries, *etc.*

(3) Under the background of economic globalization and economic transition in China, how to explore diversified paths to urbanization with regional characteristics for border areas by taking into account the restriction of natural resources, the opening up of the frontier regions, impetus of poverty alleviation, national security, and the demands for pushing local economic development?

(4) Confronted with the complex poverty problems such as urban poverty, new poverty population, poverty-returning population, and long-term poverty population, *etc.*, what kinds of poverty alleviation systems should be established in border areas under the restrictions of natural, economic, concept, technological and labor force factors?

(5) How to develop tourism industry with local characteristics in the underdeveloped border areas where the territory is vast but sparsely populated. There are restricting factors in border areas, such as the high development cost, big development difficulties and low level of local economic development, as well as various responsibilities, such as the ecological environment conservation, handing down of the culture of minority nationalities, poverty alleviation, employment, and local economic development.

(6) What kinds of security systems should be established in border areas based on the interaction and mutual influences among the aspects of geo-politics, geo-economy, geo-society, geo-culture, geo-nationalities, geo-population, *etc.*?

(7) Ecological environment in constructing harmonious society in border areas, including the influence degree of ecological environment in the headwaters and source fields on global climate change. It is the national scientific and technological demand to carry out research on the ecological compensation system for the regions where development is forbidden.

(8) The strategies of transnational regional cooperation and bidirectional economic development in border areas: how to construct the characteristic industrial clusters and the import-export processing bases with consideration on both national and local benefits, the balance between the feasibility of cost benefit and the maximization of local economic benefits, the regional advantages of water, land, sunshine and heat, the renewal of traditional agricultural technologies, and domestic and foreign market demands?

A2.4.3 Overall Goals and Stage Objectives to 2050

(A) Overall Goals

It is planned to achieve the comprehensive and systematical research achievements in nine key fields to 2050, including the exploitation of strategic resources, the opening up to the outside world, town development, tourism development, water and land exploitation, development and poverty alleviation in the minority regions, agriculture and characteristic industries, ecological environment and geo-politic security, which will provide scientific basis for the sustainable development in the border areas in China.

(B) Short-term Goals (to 2020)

The main short-term goals are to research the regional features, existing problems, impetus and resistance factors in regional development, and reveal the evolution laws of the spatial patterns of border areas; to analyze the affecting factors of future regional development, to identify the development orientation of border areas in China and international cooperation, and to study the regulatory approaches and countermeasures to the resisting factors of regional economic development in border areas; to put forward the reasonable modes of regional division of labor and cooperation between the border areas and central and eastern China and neighboring countries; to develop evaluation index systems for the stable and coordinated development of society, economy and ecological environment in border areas, and to probe into the optimal regional development modes and regulatory mechanisms of the regional characteristics of border areas; to put forward the industrial systems, spatial distribution, implementation approaches, countermeasures and policies of the key fields in the regional development of border areas as well as the coordination mechanism between the regional economic development and the ecological and political security in border areas. The emphases of the researches are as follows: 1) analysis and evaluation index systems of the status quo of border areas; 2) reasonable modes of regional division of labor and cooperation between the border areas and central and eastern China and neighboring countries; 3) industrial systems, spatial distribution, implementation approaches, countermeasures and policies of the key fields; 4) coordination mechanism between the regional economic development and the ecological and political security in border areas.

(C) Medium-term Goals (to 2030)

Based on the frontier of the discipline and significant national strategic demands, the medium-term goals are to develop the theoretical frameworks, integration methods and application systems for the rise and development in the border areas of China; to reveal the bidirectional economic development pattern in border areas, laws and dynamic mechanisms of transnational and inter-regional material, energy, information and capital flows; to elucidate the laws of regional differentiation and the approaches of urban-rural harmonious development for the regional economic development in the border areas; to study the development modes, mechanisms, and countermeasure and policy systems for regional economic integration. The emphases of the researches are as follows: 1) study on the theories of the bidirectional economic development in border areas; 2) study on the key industrial clusters and innovation systems; 3) study on the process of regional economic integration; 4) study on the ecological and environmental effects of the regional development in border areas.

(D) Long-term Goals (to 2050)

The main long-term goals are to study the reorganization of the world economic patterns based on the optimization of cost-benefit and the response of border areas under the background of new stages, new factors and new patterns; to study the security system construction, modes and mechanisms of the regional economic integration in border areas.

A2.4.4 Implementation Plans and Technological Roadmap

(A) Research Contents

(1) Research on the construction of industrial system and spatial layout of strategic resources, including the investigation and assessment of strategic resources, selection of industrial systems and optimized layout of industrial clusters, industrial layout and optimized allocation of clean energy, and innovation systems of strategic resources.

(2) Research on the system of opening to the outside world and its spatial patterns for border areas, including the comparative analysis of the border areas and the neighboring countries as well as the strategic orientation in China; construction of the bidirectional industrial systems, construction and layout of the international corridors in border areas; the modes, mechanisms and policy systems of foreign economic cooperation.

(3) Research on the transnational urban systems and their spatial patterns in border areas, including the development mechanisms, spatial patterns and laws of urban morphological evolution, urbanization process and urban-rural integration in border areas, and optimized spatial patterns of transnational urban systems.

(4) Research on the development of transnational tourism industrial systems, including the optimized modes and spatial patterns of tourism industrial development, and cooperation mechanisms and modes of transnational tourism in border areas.

(5) Research on utilization of water resources and water-land exploitation in the international river basins in border areas, including the macro-strategies for the water-land exploitation, trans-basin water diversion and international relations with the countries at the lower reaches on the basis of guaranteeing the security of cultivated land resources and the ecological security in the ecologically sensitive and fragile regions in China. To investigate and rationally allocate water and land resources; to assess the effects of trans-basin water diversion on ecological environment, and to develop comprehensive optimization and allocation technologies of water and land resources, agricultural production, and ecological and environmental conservation in

border areas; to explore the approaches of optimizing and analyzing the multi-variable and large-area regional allocation of water and land resources; to establish the decision-making models for the optimization, exploitation and technical prediction of water and soil resources in different regions.

(6) Research on the regional development and poverty alleviation systems in the minority regions, including the poverty mechanism, poverty-returning mechanism and poverty alleviation systems in border areas.

(7) Research on the improvement of traditional agriculture and the development of advantageous and characteristic industries in border areas, including the land use and agricultural development in border areas, and the optimized layout of characteristic and advantageous industrial clusters.

(8) Research on global climate change and the response mechanism of border areas, including the ecological environmental problems and ecological conservation and restoration in border areas, and the ecological environment in the headwaters and windward regions and its response to global climate change.

(9) Research on the geopolitical development strategies in border areas, including the regional cooperation mechanisms, cooperation modes, security systems and development strategies in border areas, the tradeoff-balance-controlled-balance modes and mechanisms of national relations by jointly building the free-trade areas, custom unions and transnational urban systems, reorganization of the transnational industrial clusters, construction of the international corridors, *etc*.

(10) Research on theories of the security in border areas: to study the theories of the opening up of border areas to the outside world and the security systems of border areas by referring to the domestic and foreign research achievements and by summarizing the prominent problems and experiences in the opening up of borders areas since the reform and opening up in China; to study the national relationship mechanism based on the interaction and mutual influences of geo-politics, geo-economics, geo-society, geo-culture, geo-nationalities, geo-population, *etc*.

(B) Research Scheme

After comprehensively analyzing and summarizing the available research achievements, a database of related spatial, social and economic information will be developed and explored through comprehensive and integrated approaches of combining qualitative and quantitative methods, scientific theories and experts' judgments with the help of GIS technology. The existing problems in security and regional development will be systematically analyzed, and the general laws of regional development and security in border areas will be summarized. In view of the uniqueness of the opening up of border areas in western China, new ideas, new approaches and new modes will be put forward to accelerate the opening up of border areas to the outside world and promote the coordinated social, economic and environmental development through analyzing the regional development and the dynamic mechanism of security in the border areas in China. Moreover, a preliminary scheme of opening up and regional development will be developed by referring to the research achievements of "Major Function Oriented Zoning".

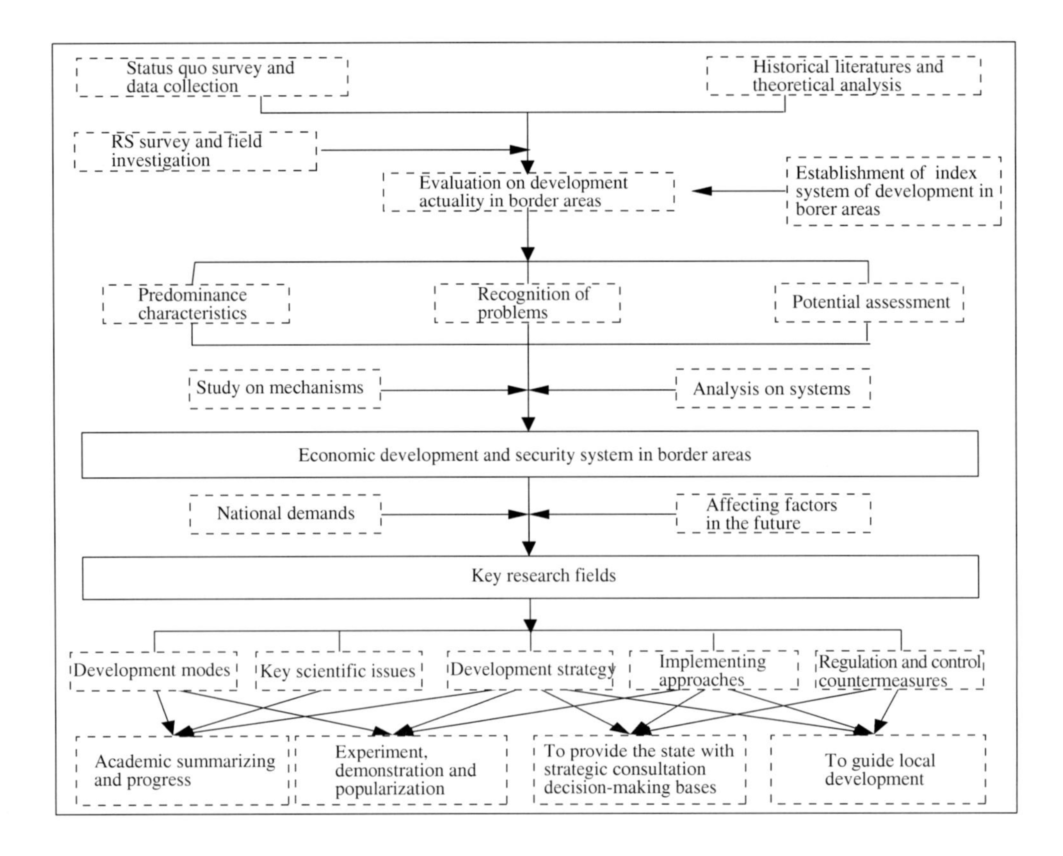

App. Fig. 2.3 Technical roadmap for research on regional development of border areas

(C) Roadmap Design

The importance degree and development focuses of different fields in the regional development of border areas are varied in different stages of development. The detailed technological route is presented as shown in App.Fig. 2.4.

(D) Suggestions and Countermeasures of Chinese Academy of Sciences

(1) To develop long-term and effective high-level conversation mechanisms between the Chinese Academy of Sciences and the academies of sciences of the neighboring countries.

(2) To continuously carry out important research oriented projects and momentous research projects in border areas, and conduct international cooperative research projects on issues of common interests and common concern together with the academies of sciences of the neighboring countries.

(3) To organize the related institutes of all the academies of sciences of the neighboring countries to apply international research projects of UNESCO, World Bank, Asian Bank, *etc*.

(4) To constantly support the training and exchange programs of qualified scientists and technicians, such as the doctors and visiting scholars between the Chinese Academy of Sciences and the academies of sciences of the neighboring

countries.

(5) The Chinese Academy of Sciences and the academies of sciences of the neighboring countries co-build laboratories, research stations, research and development centers, and central scientific and technological cooperation platforms to transfer, transform and popularize the research achievements, co-publish academic journals, strengthen the exchange of scientific and technological information, and support various international conferences.

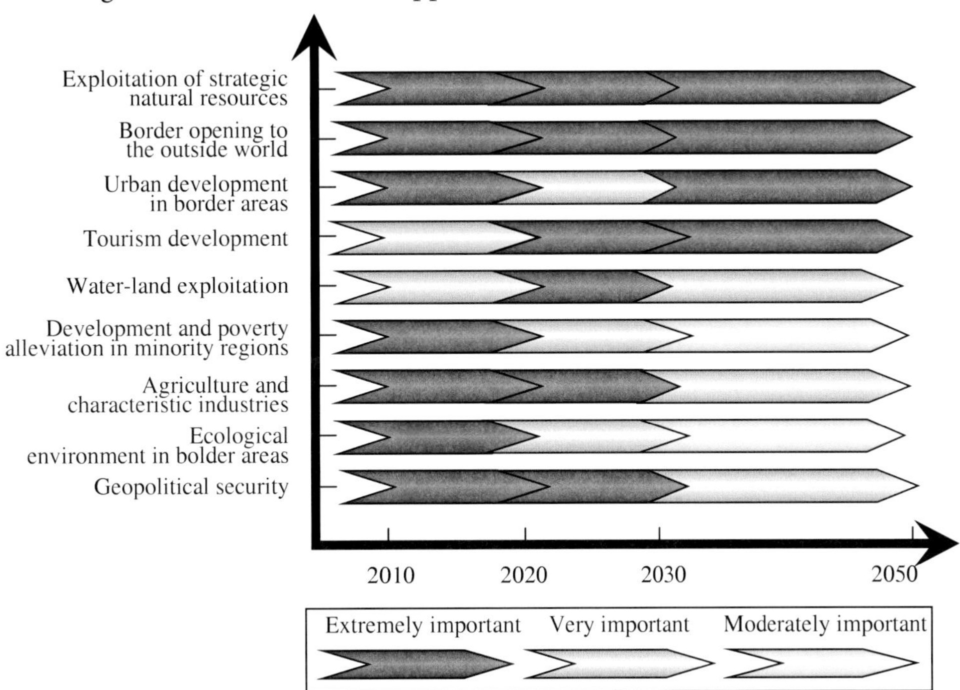

App. Fig. 2.4 Scientific and technological development roadmap for research on regional development of border areas to 2050

References

[1] Lu Dadao. Regional Development and Its Spatial Structure. Beijing: Science Press,1995. (in Chinese)

[2] Lu Dadao. Several issues regarding China's regional development strategies and policies. Economic Geography, 2009, 29 (1):2–7.(in Chinese)

[3] Lu Dadao, Yao Shimou. A scientific thought about urbanization progress in China. Human Geography, 2007, 4:1–5.(in Chinese)

[4] Healey P. The treatment of space and place in the new strategic spatial planning in Europe. International Journal of Urban and Regional Research, 2004, 28 (1):45–67.

[5] Balestra C F. Redefining Territories: The Functional Regions. Organization for Economic Co-operation and Development, 2002.

[6] Zhao Shidong, Zhang Yongming, Lai Pengfei. Millennium Ecosystem Assessment Report (1, 2). Beijing: China Environmental Science Press, 2007.(in Chinese)

[7] Lu Dadao. The Development and Innovation of Geography. Beijing: Science Press, 1999. (in Chinese)

[8] Wu Chuanjun. Man-Land Relationship and Economic Layout. Beijing: Academy Press, 2008. (in Chinese)

[9] Fan Jie. Comprehensiveness of geography and integrated research on regional development. Acta Geographica Sinica, 2004, 59 (Supp.1):33–40.(in Chinese)

[10] Fan Jie. Economic Globalization and Regional Development. Beijing: People's Education Press, 2002.(in Chinese)

[11] Zheng Du. A study on the regionality and regional differentiation of geography. Geographical Research, 1998, 17 (1):4–9.(in Chinese)

[12] Lu Dadao. New factors and new patterns of regional development in China. Geographical Research, 2003, 22 (3):261–271.(in Chinese)

[13] Fan Jie. The scientific foundation of major function oriented zoning in China. Acta Geographica Sinica, 2007, 62 (4):339–350.(in Chinese)

[14] Lu Dadao, Liuyi, Fanjie, *et al.* Chinese Regional Development Report (2002, 2000, 1999, 1997). Beijing: The Commercial Press.(in Chinese)

[15] Gu Chaolin, Zhao Lingxun, Zhang wei. China's High-tech Industries and Park. Beijing: China Citic Press, 1998.(in Chinese)

[16] Hu Xuwei, Zhou Yixing, Gu Chaolin, *et al.* Studies on the Spatial Agglomeration and Dispersion in China's Coastal City-and-Town Concentrated Areas. Beijing: Science Press, 2000.(in Chinese)

[17] Isard W. Methods of Regional Analysis: An Introduction to Regional Science. Cambridge, Mass: The MIT Press, 1967.

[18] Liu Hui, Fan Jie, Wang Chuansheng. Progress and enlightenment of european spatial planning research. Geographical Research, 2008, 27(6): 1381–1389.(in Chinese)

[19] Fan Jie. Social demands and new propositions of economic geography discipline development based on the Eleventh National Five-Year Plan. Economic Geography, 2006, 26(4):545–550. (in Chinese)

[20] Lu Dadao. Development of economic geography in China. World Regional Studies, 2007, 16 (4):1–4.(in Chinese)

[21] Liu Weidong, Liu Yansui, Jin Fengjun,*et al.* China's Regional Development Report. Beijing: The Commercial Press, 2007.(in Chinese)

[22] Niu Wenyuan. The Pandect of Sustainable Development in China. Beijing: Science Press, 2007. (in Chinese)

[23] Kuznets, Simon. Economic Growth of Nations. Beijing: Commercial Press, 2007.(in Chinese)

[24] Wei Houkai. Market Competition, Economic Performance, and Industrial Concentration: An Empirical Study on China's Manufacturing Concentration and Market Structure. Beijing: Economy & Management Publishing House, 2003.(in Chinese)

[25] Cai Jianming, Wang Guoxia, Yang Zhenshan. Future trends and spatial patterns of migration in China. Population Research, 2007, 31(5):9–19.

[26] The Study Group of National Population Development Strategy.The Report of National Population Development Strategy. Beijing: China Population Press, 2008.(in Chinese)

[27] Zhou Yixing. Thoughts on the speed of China's urbanization. Urban Planning, 2006,(Supp.1): 32–35.(in Chinese)

[28] Saskia Sussen. The Global City: New York, London, Ttokyo. New Jersey: Princeton University Press, 2001.

[29] Sun Honglie, Zhang Rongzu. The Principle and Experience of Zonality in China's Ecological Environment Construction. Beijing: Science Press, 2004.(in Chinese)

[30] Li Wenyan. Several issues on China's energy strategy in the early 21st century. Economic Geography, 2000, 20(1):7–12.(in Chinese)

[31] Liang Jinshe, Hong Lixuan, Cai Jianming. The decomposition of energy consumption growth–During the process of China's urbanization: 1985–2006. Journal of Natural Resources, 2009, 24 (1):20–29.(in Chinese)

[32] Zhang Wenchang, Jin Fengjun, Fan Jie. Traffic Economic Belt. Beijing: Science Press, 2000. (in Chinese)

[33] Jin Fengjun, Wang Chengjin, Li Xiuwei. Discrimination method and its application analysis of regional transport superiority. Acta Geographica Sinica, 2008, 63(8):787–798.(in Chinese)

[34] Liu Weidong, Dicken P, Yang Weicong. The impacts of new information and communication technologies on the spatial organization of firms: a case study of the Xingwang Industrial Park in Beijing. Geographical Research, 2004, 23 (6):833–844.(in Chinese)

[35] Gu Chaolin, Zhang Qin, Cai Jianming, *et al*. Economic Globalization and China's Urban Development. Beijing: The Commercial Press, 1999.(in Chinese)

[36] Ma Li, Liu Weidong, Liu Yi. Spatial evolution of local production network under economic globalization. Geographical Research, 2004, 23(1):87–96.(in Chinese)

[37] Lu Dadao, Liu Yi, Fan Jie. Policy effects and status of China's regional development. Acta Geographica Sinica, 1999, 54 (6):496–508.(in Chinese)

[38] Lu Dadao. Theories and Experiences of China's Regional Development. Beijing: Science Press, 2003.(in Chinese)

[39] Wu Chuanjun. Geography. Fuzhou: Fujian Education Press, 2000.(in Chinese)

[40] Lu Dadao. Science Development and China's approach to urbanization. Academic Journal of Suzhou University (Philosophy & Social Science), 2007, 28 (2):1–7.(in Chinese)

[41] Guo Huancheng. Agricultural Economy Zoning in China. Beijing: Science Press, 1999.(in Chinese)

[42] Hou Xueyu. Natural Ecological Zoning and Development Strategies of General Agriculture in China. Beijing: Science Press, 1988.(in Chinese)

[43] Huang Bingwei. A draft of comprehensive natural zoning of China. Science Aviso, 1959, (18): 594–602.(in Chinese)

[44] Zhou Lisan. Comprehensive Agricultural Zoning of China. Beijing: China Agriculture Press, 1981. (in Chinese)

[45] Zheng Du, Ge Quansheng, Zhang Xueqin, *et al*. Regionalization in China: retrospect and prospect. Geographical Research, 2005, 24 (3):330–344.(in Chinese)

[46] Yang Qinye, Wu Shaohong, Zheng Du. A retrospect and prospect of researches on regional physio-geographical system (RPGS). Geographical Research, 2002, 21(4):407–417.(in Chinese)

[47] Clark D. Urban World/ Global City. London: Routledge, 1996.

[48] Zhou Yixing. Speed of urbanization: not the faster the better. Scientific Decision-Making Magazine,

2005, (8):30–33

[49] Ding Sibao, Wang Xiaoyun. Theory and mechanism of the regional eco-environmental compensation. Journal of Northeast Normal University (Philosophy & Social Sciences), 2008, (4):5–10.(in Chinese)

[50] Wang Rusong, Hu Dan. Implementation ecological civilization and promoting development of ecological sciences. Acta Ecologica Sinica, 2009, 29(3): 1055–1067.(in Chinese)

[51] Wei Houkai, Liu Kai. The analysis and prediction on the trend of China's regional differences. China Industrial Economics, 1994, (3):28–36.(in Chinese)

[52] Qin Yanhong, Kang Muyi. A review of ecological compensation and its improvement measures. Journal of Natural Resources, 2007, 22 (4): 557–567.(in Chinese)

[53] Lai Li, Huang XianJin, Liu WeiLiang. Advances in theory and methodology of ecological compensation. Acta Ecologica Sinica, 2008, 28 (6):2870–2877.(in Chinese)

[54] Yang Guangmei, Min Qinwen,Li Wenhua, *et al*. Scientific issues of ecological compensation research in China. Acta Ecologica Sinica, 2007, 27 (10):4289–4300.(in Chinese)

[55] Ye Wenhu, Wei Bin, Tong Chuan. Measurement and application of urban ecological compensation. China Environmental Science, 1998, 18(4):298–301.(in Chinese)

[56] Zhao Jingzhu, Luo Qishan, Yan Yan, *et al*. Thoughts on improving China's ecological compensation mechanism. Macroeconomic Management, 2006, (8):53–54.(in Chinese)

[57] Wang Rusong, Ouyang Zhiyun. Some considerations with scientific views on ecological security in China. Bulletin of Chinese Academy of Sciences, 2007, 22 (3): 223–229.(in Chinese)

[58] Hanson S. The Geographic Ideas that Changed the World. New Jersey: Rutgers University Press, 1997: 8–9.

[59] Lu Dadao. Location Theory and Regional Research Methods. Beijing: Science Press, 1988: 14–47. (in Chinese)

[60] Li Xiaojian.Economic Geography. Beijing: Higher Education Press, 1999.(in Chinese)

[61] Gibs D, Healey M. 1997. Industrial geography and the environment. Applied Geography, 17(3): 193–201.

[62] Angel D. 2000. Environmental innovation and regulation. In: Clark F, Gertler (eds.). The Oxford Textbook of Economic Geography. Oxford: Oxford University Press, 2000: 607–622.

[63] Climate Committee of Geographical Society of China. Global change research and the geography. In: Wu Chuanjun, *et al*. (eds.). Chinese Geography at the Turn of the Century. Beijing: People's Education Press, 1999: 68–76.(in Chinese)

[64] Sneddon C S. "Sustainability" in ecological economics, ecology and livelihoods: a review. Progress in Human Geography, 2000, 24(4):521–549.

[65] Wu Chuanjun.Chinese Economic Geography. Beijing: Science Press, 1998: 256–257.(in Chinese)

[66] Li Yining. New Thoughts of Regional Development. Beijing: Economic Daily Press, 2000. (in Chinese)

[67] Hu Xuwei, Mao Hanying, Lu Dadao. Problems and countermeasures of sustainable development in China's coastal areas. Acta Geographica Sinica, 1995, 50(1):1–12. (in Chinese)

[68] The Geographical Society of China. Research on Regional Sustainable Development. Beijing: China Environmental Science Press, 1997.(in Chinese)

[69] Lang Yihuan, Wang Limao. The types of deficient resources and the trend of supply and demand. Journal of Natural Resources, 2002, 17(4): 409–414.(in Chinese)

[70] Li Xiaojian. Foreign direct investment and its impact on economic development of coastal China. Acta Geographica Sinica, 1999, 54 (5): 420–430.(in Chinese)

[71] Ge Quansheng, Fang Xiuqi, Zhang Xueqin, *et al*. Remarkable environmental changes in China during the past 50 years: a case study on regional research of global environmental change. Geographical Research, 2005, 24 (3):345–358.(in Chinese)

[72] Yang Guiqing. Ten major affecting factors of American metropolitan development in the next fifty

years and considerations on current chinese urban development: introduction of Robert Fishman's "The American Metropolis at Century's End". Urban Planning Forum, 2006, (5): 103–110. (in Chinese)
[73] Wang Mingfeng. Cultural industry policy and urban development: the European experience and enlightenment. Urban Studies, 2001, (4):11–15.(in Chinese)
[74] Yang Daobing, Lu Jiehua. Analysis on aging labor force and its impact on socio-economic development in China. Population Journal, 2006, (1):7–12.(in Chinese)
[75] Gu Shuzhong, Geng Haiqing, Yao Yulong. The compartmentalization of functional areas for national resource security and the orientation of the Western China. Progress in Geography, 2002, (5): 410–419.(in Chinese)
[76] Shen Lei, Liu Xiaojie. Discussion on theories and methods of resources flow. Resources Science, 2006, 28 (3):9–16.(in Chinese)
[77] Zhang Wenzhong. Theories and Experience of Industrial Development and Planning. Beijing: Science Press, 2009.(in Chinese)
[78] Zhen Feng. Regional development strategy and regional planning under information era. Urban Planning Forum, 2001, (6):61–64.(in Chinese)
[79] Li Wenyan, Ni Zubin. Economic work in comprehensive resource investigation over the past 30 years. Paper for 1979 Comprehensive Conference of Geographical Society of China.
[80] Li Wenyan. The role of natural factors and technological-economic factors in industrial allocation. Acta Geographica Sinica, 1957, 23 (4): 399–417.(in Chinese)
[81] Hu Xuwei, Hu Tingzhu. Tech-economic evaluation on industrial allocation. Acta Geographica Sinica, 1965, 31 (3):179–193.(in Chinese)
[82] Zhang Wenchang. Issues on the allocation of transportation in industrial bases. Acta Geographica Sinica, 1981, 36 (2): 157–170.(in Chinese)
[83] Hu Zhaoliang. The tendency in the industrial distribution in China. Acta Geographica Sinica, 1986, 41(3): 193–201.(in Chinese)
[84] Li Wenyan. Regional differentiation of mineral resources and geographical locations in China—An economic geographical analysis of several conditions of industrial distribution. Geographical Research, 1982, 1(1):19–30.(in Chinese)
[85] Li Wenyan. Industrial Geography in China. Beijing: Science Press, 1990.(in Chinese)
[86] Lu Dadao. Patterns of grouping firms in industrial districts and their techno-economic effects. Acta Geographica Sinica, 1979, 34 (3): 248–264.(in Chinese)
[87] Shen Xiaoping. A discussion on the economic effects of grouping firms in industrial districts and determination of the optimal size. Acta Geographica Sinica, 1987, 42 (1): 51–61.(in Chinese)
[88] Pang Xiaomin. The changing industrial location factors and the new trends of industrial geography research. Geographical Research, 1992, 11(3):101–107.(in Chinese)
[89] Lu Dadao. Theories and Experience of Industrial Allocation in China. Beijing: Science Press, 1990. (in Chinese)
[90] Lu Dadao. Macro strategy of regional development in China. Acta Geographica Sinica, 1987, 42(2):97–105.(in Chinese)
[91] Li Xiaojian. New industrial district and globalization: A literature review. Progress in Geography, 1997, 16(3): 16–23.(in Chinese)
[92] Wang Jici. Modern Industrial Geography. Beijing: China Science and Technology Press, 1994. (in Chinese)
[93] Wei Xinzhen, Wang Jici. New Industrial Space: the Development and Layout of High-tech Industry Development Area. Beijing: Peking University Press, 1993.(in Chinese)
[94] Liu Weidong, Xue Fengxuan. The changing spatial organization of the automotive industry: the impact of production pattern changes. Progress in Geography, 1998, 17 (2):1–14.(in Chinese)
[95] Dicken P. Global Shift: Reshaping the Global Economic Map in the 21st Century. New York: The

Guilford Press, 2003.
[96] Yeung H W. Organizing "the firm" in industrial geography I: networks, institutions and regional development. Progress in Human Geography, 2000, 24 (2): 301–315.
[97] Yeung H W. Business networks and transnational corporations: a study of Hong Kong firms in the ASEAN Region. Economic Geography, 1997, 73 (1): 1–25.
[98] Dicken P, Philip F K, Olds K, Yeung H W. Chains and networks, territories and scales: toward a relational framework for analyzing the global economy. Global Networks, 2001, 2 (1):89–112.
[99] Institute of Industrial Economics, CASS. China Industry Development Report: 2007. Beijing: Economy & Management Publishing House, 2007.(in Chinese)
[100] Zhang Xiaolin. On discrimination of rural definitions. Acta Geographica Sinica, 1998, 53 (4): 365–371. (in Chinese)
[101] Guo Huancheng. Nature and tasks of rural geography. Economic Geography, 1988, 8 (2): 125–129. (in Chinese)
[102] Zeng Zungu. Academician Wu Chuanjun and agricultural geography research. Areal Research and Development, 2008, 27(2): 6–7.(in Chinese)
[103] Deng Jingzhong. Preliminary study on regionalization of Chinese status quo agriculture. Acta Geographica Sinica, 1963, 29 (4): 265–280.(in Chinese)
[104] Deng Jingzhong. Some problems on the comprehensive agricultural regionalization of China. Geographical Research, 1982, 1 (1): 9–18.(in Chinese)
[105] Chen Shupeng, Yang Lipu. The settlement around Zunyi. Acta Geographica Sinica, 1943, 10 (1): 69–81.(in Chinese)
[106] Jin Qiming. The history and current trends of research on rural settlement geography in China. Acta Geographica Sinica, 1988, 43 (4): 311–317.(in Chinese)
[107] Fei Xiaotong. Peasant Life in China: A Field Study of County Life in the Yangtze Vally. Beijing: The Commercial Press, 2006.(in Chinese)
[108] Zhou Lisan. Agricultural Geography in China. Beijing: Science Press, 2000.(in Chinese)
[109] Chen Guojie. Development of West China and the construction of settlement ecology—Take the southwest mountainous area of China as an example. Rural Eco-environment, 2001, 17(2): 5–8. (in Chinese)
[110] Liu Yansui, Lu Dadao. The basic trend and regional effect of agricultural structure adjustment in China. Acta Geographica Sinica, 2003, 58 (3): 381–389.(in Chinese)
[111] Cai Y L, Smit B. Sustainability in agriculture: a general review. Agriculture, Ecosystems and Environment, 1994, 49 (3), 299–307.
[112] Cai Y L, Smit B. Sustainability in Chinese agriculture: challenge and hope. Agriculture, Ecosystems and Environment, 1994, 49 (3), 279–288.
[113] Huang J K, Rozelle S, Zhang L X. WTO and Agriculture: Radical reforms or the continuation of gradual transition. China Economic Review, 2001, 11: 397–401.
[114] Wu Chuanjun. Sustainable Development Issues of China's Agriculture and Rural Economy: Empirical Study on Different Types of Regions. Beijing: China Environmental Science Press, 2001. (in Chinese)
[115] Wu C J. The new development of rural China. BEVAS-SOBEG, 1997, 1: 101–105.
[116] Huang Jikun. Development and prospect of Chinese agriculture. Management Review, 2003, 15 (1): 17–20.(in Chinese)
[117] Li Peilin. Judgments on current social contradictions. Outlook, 2004, (2): 61–61.(in Chinese)
[118] Zhang L X, Rozelle S, Huang J K. Off-farm Jobs and on-farm work in periods of boom and bust in rural China. Journal of Comparative Economics, 2001, 29 (3): 505–526.
[119] Chen Xiwen. Deepen rural reform and promote urban-rural overall planning. Management and Administration on Rural Cooperative, 2008, 5: 1.(in Chinese)
[120] Li Zhensheng. Area preserving, unit yield promoting and consumption saving: reflections on

China's grain production and consumption. Qiushi, 2008, 6: 44–45.(in Chinese)
[121] Han J. Building a new countryside: a long-term task in China's modernization drive. China Economist, 2007, (6): 93–111.
[122] Liu Yansui. The research progress of new concepts and modes of new rural construction. Geographical Research, 2008, 22 (2): 479–480.(in Chinese)
[123] Long H L, Liu Y S, Li X B, Chen Y F. Building new countryside in China: a geographical perspective. Land Use Policy, 2010. (in press)
[124] Liu Y S, Wang L J, Long H L. Spatio-temporal analysis of land-use conversion in the eastern coastal China during 1996–2005. Journal of Geographical Sciences, 2008, 18 (3): 274–282.
[125] Long H L, Liu Y S, Wu X Q, Dong G H. Spatio-temporal dynamic patterns of farmland and rural settlements in Su-Xi-Chang Region: implications for building a new countryside in coastal China. Land Use Policy, 2009, 26 (2): 322–333.
[126] Long H L, Zou J, Liu Y S. Differentiation of rural development driven by industrialization and urbanization in eastern coastal China. Habitat International, 2009, 33 (4): 454–462.
[127] Qiao Jiajun. Theory of Territorial Economy of China's Rural Areas. Beijing: Science Press, 2008. (in Chinese)
[128] Antrop M. Landscape change and the urbanization process in Europe. Landscape and Urban Planning, 2004, 67 (1–4): 9-26.
[129] Chen Weibang. The healthy development of urbanization is the important guarantee of harmonious society. Journal of Shanghai Polytechnic College of Urban Management, 2006, 22 (1): 1–2. (in Chinese)
[130] Qiu Baoxing. Challenges faced by China in its rapid urbanization process in the near future. Urban Studies, 2003, 21 (6): 12–14.(in Chinese)
[131] National Bureau of Statistics of China. China Statistical Abstract 2008. Beijing: China Statistics Press, 2008: 32–34.(in Chinese)
[132] Fang Chuanglin. The urbanization and urban development in China after the reform and opening-up. Economic Geography, 2009, 1: 25–31.(in Chinese)
[133] Lu Dadao, Yao Shimou, Liu Hui, *et al*. Chinese Regional Development Report: Urban Development and Spatial Expansion. Beijing: The Commercial Press, 2007: 12–19.(in Chinese)
[134] Fang Chuanglin, Liu Haiyan. The spatial privation and the corresponding controlling paths in China's urbanization process. Acta Geographica Sinica, 2007, 62 (8):849–860.(in Chinese)
[135] Fang Chuanglin. A Report on China's Urbanization Progress and Its Resource and Environment Support. Beijing: Science Press, 2009: 10–25.(in Chinese)
[136] Gu Chaolin, Yu Taofang, Li Wangming, *et al*. China's Urbanization: Pattern, Process and Mechanism. Beijing: Science Press, 2008: 705–712.(in Chinese)
[137] Jackson E L, Thomas Q, Burton L. Leisure Studies: Prospects for the Twenty- First Century. State College, PA: Venture Publishing Inc, 1999.
[138] Harper, W. The future of leisure: making leisure work. Leisure Studies, 1997, (16):189–198.
[139] Zhou Jue. The Economic Analysis of Leisure. Beijing: Economic Science Press, 2007.(in Chinese)
[140] Lou Jiajun. An Exploratory discussion on leisure development trends and some countermeasures. Tourism Science, 2004, 18 (3): 45–48.(in Chinese)
[141] Wang Qiyan. Ten trends of tourism consumption in leisure era. Tourism Tribune, 2006, 21(10):7–9. (in Chinese)
[142] Wei Xiaoan. On developing the leisure industry. Journal of Zhejiang University (Humanities and Social Sciences), 2006, 36 (5):107–114.(in Chinese)
[143] Li Zhongguang, Lu Changchong. Fundamental Leisure Studies. Beijing: Social Sciences Academic Press, 2004.(in Chinese)
[144] Chen Hang, Zhang Wenchang, Jin Fengjun. Geography of Transportation in China. Beijing: Science Press, 2007.(in Chinese)

[145] China Association for Science and Technology. Report on Advances in Geographical Sciences: 2006–2007. Beijing: China Science and Technology Press, 2007.(in Chinese)

[146] Wang Chengjin, Jin Fengjun. Research history and developing trend about geography of transportation in China. Progress in Geography, 2005, 24 (6):66–78.(in Chinese)

[147] Zhang Wenchang, Wang Jiaoe. Changes on layout of transportation and traffic in China since reforming and opening-up. Economic Geography, 2008, 28 (5):705–710.(in Chinese)

[148] Wang Chengjin. Research prospect and progress of modern port geography. Progress in Geography, 2008, 23 (3):243–251.(in Chinese)

[149] Cao Xiaoshu, Xue Desheng, Yan Xiaopei. Development tendency of urban transport geography. Scientia Geographica Sinica, 2006, 26 (1):111–117.(in Chinese)

[150] Wang Shuhua, Mao Hanying, Wang Zhongjing. Progress in research of ecological footprint all over the World. Journal of Natural Resources, 2002, 17(6): 776–782.(in Chinese)

[151] Xu Xianli, Cai Yumei, Zhang Keli, *et. al*. Study on dynamic change of cultivated land resources and causes of the changes. China Population, Resources and Environment, 2005, 15 (3):75–79. (in Chinese)

[152] Xia Jun, Su Renqiong, He Xiwu. Water resources problems in China and their countermeasures and suggestions. Bulletin of the Chinese Academy of Sciences, 2008, 23 (2):116–120.(in Chinese)

[153] Wang Wenjun, Li Shuqing. Building circular economy to solve environment issue in the condensation model industrialization society. China Industrial Economics, 2004, (9):29–35. (in Chinese)

[154] Lu Dadao,Yao Shimou, Li Guoping, *et al*. Comprehensive analysis of the urbanization process based on China's conditions. Economic Geography, 2007 (6): 883–887.(in Chinese)

[155] Liu Lili. A study of strategies for the development of China's oil industry. Journal of the University of Petroleum, China (Social Sciences), 2004, 20 (1): 1–6. (in Chinese)

[156] Xu Liang. Considerations on national energy strategy of China. China Soft Science, 2006, (7):29–32. (in Chinese)

[157] Tan Minghong, Li Xiubin, Lv Changhe. Construction land expansion and its occupancy of cultivated land in China's large and medium-sized cities in the 1990s. Science in China (Ser.D), 2004, 34 (12):1157–1165.(in Chinese)

[158] Tian Guangjin, Liu Jiyuan, Zhuang Dafang. The temporal-spatial characteristics of rural residential land in China in the 1990s. Acta Geographica Sinica, 2003, 58 (5):651–658.(in Chinese)

[159] Fu Qingyun. Structural change and development direction of energy sources in the United States, Germany, Britain and Japan. Land and Resources Infromation, 2005, (7):8–12.(in Chinese)

[160] Cao Xin. Energy saving and emissions reducting in power industry and countermeasures. China National Conditions and Strength, 2008, (4): 39–41.(in Chinese)

[161] Sun Qiang, Cai Yunlong. Historical experiences of cropland conservation and land management in Japan. Acta Scientiarum Naturalium Universitatis Pekinensis, 2008, 44 (2):249–256.(in Chinese)

[162] Yu Bohua. The change of arable land area in Japan since the 1960s and its policy implications. Resources Science, 2007, 29 (5): 182–189.(in Chinese)

[163] .Jiang Yihua. Water resources management in Japan and its enlightenment. Economic Research Guide, 2008, 37 (18):180–183.(in Chinese)

[164] Wei Yaping, Wang Jiwu. Urban expansion and urban and town sprawl: policy analysis of land-use regulations in the urban growth of China's mega-cities. China Land Science, 2008, 22 (4):19–24. (in Chinese)

[165] Liang Shumin. Research on China's urbanization development based on cultivated land preservation strategies. China Land Science, 2009, 23 (5):41–46.(in Chinese)

[166] Dong Suocheng. Report of China's Resources, Environment and Development in a Century. Wuhan: Hubei Science and Technology Press, 2002.(in Chinese)

[167] Cai Yumei, Zhang Wenxin, Liu Yansui. Forecasting and analyzing the cultivated land demand

based on multi-objectives in China. Resources Science, 2007, 29 (4):134–138.(in Chinese)

[168] Yang Hongshan. Institutional analysis on digital urban management. Urban Studies, 2009, (1): 109–113.(in Chinese)

[169] Zhang Jingxiang. Urban and regional governance and the research and application of it in China. Urban Problems, 2000, 98 (2): 40–44.(in Chinese)

[170] Qian Zhenming. Urban management: a transition from tradition to modernization. Chinese Public Administration, 2005, 237(3): 37–40.(in Chinese)

[171] Li Ming, Fang Chuanglin, Sun Xinliang. Progress and prospect in regional governance study. Progress in Geography, 2007.

[172] Yi Xudong, Dai Xuan. Development and changes of China's regional management in economic globalization. Productivity Research, 2003, (5): 136–137.(in Chinese)

[173] Fan Jie. On territorial (regional) planning in the new era and its theoretical foundation construction. Progress in Geography, 1998, 17 (4):1–7.(in Chinese)

[174] Gareth D M. Public Economics. New York : Cambridge University Press, 1995.

[175] Li Chuncheng. Research on Controlled Resources by the Government and Regulation Methods. Chengdu: Sichuan University Press, 2002.(in Chinese)

[176] Clark G L, Feldman M P, Gertler M S. The Oxford Handbook of Economic Geography. Beijing: The Commercial Press, 2005.(in Chinese)

[177] Zhang Fanrong, Xue Xiongzhi. On the construction of collaboration patterns between local government in integrated regional marine management. Ocean Development and Management, 2009, 26 (1):21–25.(in Chinese)

[178] Research Group of China's Development Strategy of Science and Technology. Annual Report of Regional Innovation Capability of China. Beijing: Science Press, 2006.(in Chinese)

[179] Yao Shimou, Chen Zhenguang, Zhu Yingming. Urban Agglomerations in China. HeFei: University of Science and Technology of China Press, 2006.(in Chinese)

[180] He Canfei, Xie Xiuzhen. Geographical cconcentration and provincial specialization of Chinese manufacturing industries. Acta Geographica Sinica, 2006, 61 (2): 212–222.(in Chinese)

[181] Han Zenglin. Distribution and optimization of container transportation network in China. Acta Geographica Sinica, 2002, 57 (4): 480–488.(in Chinese)

[182] She Zhixiang, Luo Yongming. Water-Soil Resource and Environment and Sustainability of the Yangtze River Delta. Beijing: Science Press, 2007.(in Chinese)

[183] Yang Guishan. Environmental Changes in Coastal China and the Regional Response. Beijing: Higher Education Press, 2001.(in Chinese)

[184] China Association for Science and Technology. Comprehensive Report on Subject Development of Resources Science. Beijing: China Science and Technology Press, 2007.(in Chinese)

[185] Jin Fengjun, Zhang Pingyu, Fan Jie, *et al*. Research on Revitalization of Northeast China and Sustainable Development Strategy. Beijing: The Commercial Press, 2006.(in Chinese)

[186] Hudson R. Institutional change, cultural transformation and economic regeneration: myths and realities from Europe's old industrial region. In: Amin A, Thrift N(eds.). Globalization, Institutions and Regional Development in Europe. Oxford: Oxford University Press, 1994: 331–345.

[187] Hassink R, Shin Dong-Ho. The restructuring of old industrial areas in Europe and Asia. Environment and Planning A, 2005, 37 (4):571–580

[188] Zhang Pingyu. Regional Development Report of Northeast China 2008. Beijing: Science Press, 2008.(in Chinese)

[189] Zhang Guobao. Research on the Planning of Revitalizing Northeast China (Volume of Comprehensive Planning Research). Beijing: Standards Press of China, 2008.(in Chinese)

[190] Fan Jie, Sun Wei, Fu Xiaofeng. Problems, reasons and strategies for sustainable development of mining cities in China. Journal of Natural Resources, 2005, 20 (1): 68–77.(in Chinese)

[191] Zhang Pingyu. Study on global environmental change and the role of human geography. World

Regional Studies, 2007, 16 (4): 76–81.(in Chinese)
[192] Wei Houkai. Effect evaluation and adjustment approach of revitalization policies for Northeast China. Social Science Journal, 2008, (1): 60–65.(in Chinese)
[193] Chen Guojie. Report on Development of China's Mountainous Areas in 2003. Beijing: The Commercial Press, 2004.(in Chinese)
[194] Deng Wei, *et al*. The conception of mountain science development in China. Bulletin of the Chinese Academy of Sciences, 2008, 23 (2):156–161.(in Chinese)
[195] Chen Guojie, Fang Yiping, Chen Yong, *et al*. Report on Development of China's Mountainous Areas: Research on Rural Settlements in China's Mountainous Areas. Beijing: The Commercial Press, 2007.(in Chinese)
[196] Deng Wei, Fang Yiping, Liu Shaoquan. On the status quo and development trend of mountain resources research. In: The Chinese Association for Science and Technology. Report on Advances in Resources Science (2008–2009). Beijing: China Science and Technology Press, 2009: 88–109. (in Chinese)
[197] Chen Guojie. Development situation of Chinese mountainous areas in recent years and strategic prospect. Bulletin of the Chinese Academy of Sciences, 2008, 23 (6) : 485–491.(in Chinese)
[198] Chen Guojie. Some considerations on strategy of development of mountain regions of China. Bulletin of the Chinese Academy of Sciences, 2007, 22 (2):126–131.(in Chinese)
[199] Zhou Wancun, Jiang Xiaobo. Developing trace of GIS and design of digital mountain. Journal of Mountain Science, 2006, 24 (5): 620–627.(in Chinese)
[200] Wang Guolian. The security environment around China and relative geopolitical strategies. World Regional Studies, 2003, (2): 99–105. (in Chinese)
[201] Shen Weilie, Lu Junyuan. Geography of China's National Security. Beijing: Shi Shi Publishing House, 2001: 328.(in Chinese)
[202] Chen Guojie. An analysis on the cause of development gaps between the East, Centre and West of China. Scientia Geographyica Sinica, 1997, 17 (1):1–7.(in Chinese)
[203] An Husen. Industrial distribution, welfare difference and government regulation. Social Sciences in Guangdong, 2007, (4):33–41.(in Chinese)
[204] Lu Mingyuan. Discussion on the functions of NGO in the regional harmonious Development. Urban Problems, 2007, (9):74–78.(in Chinese)

Epilogue

In recent years, the regional development research team at the Chinese Academy of Sciences has excellently accomplished a series of scientific research missions on significant issues of regional sustainable development in China by bringing the interdisciplinary advantages into play, which has won the recognition of the government and academia, and CAS' influence as a "think tank" for national strategic decision-making has thus been expanded. The research on the "Roadmap of the Development of Science and Technology in China's Regional Development Research During the Additional Years to 2050" ("Roadmap of Regional Development" for short) will undoubtedly push the regional development research of CAS into a new stage of development.

The project of "Roadmap of Regional Development" started at the end of 2007. According to the overall project design, the research has been divided into three parts:

The first part is the general study which explores the characteristics of regional development research, future trends and scientific and technological demands of regional development, the significant propositions of regional development, and the strategies of CAS in this field with Chapter 1, 2, 3, 4, 5 written by Dadao Lu and Jie Fan.

The second part is monographic studies by the combined efforts of middle-aged and young scholars. These monographic studies of the roadmap focus on the important research fields or sectors of regional development, including new factors in regional development (Appendix 1: section 1, written by Weidong Liu and Zhiyong Hu), industry (Appendix 1: section 2, written by Wenzhong Zhang and Li Ma), agriculture and countryside (Appendix 1: section 3, written by Yansui Liu and Hualou Long), urbanization and population (Appendix 1: section 4, written by Chuanglin Fang and Jinchuan Huang), leisure (Appendix 1: section 5, written by Tian Chen, Jiaming Liu and Linsheng Zhong), transport infrastructure (Appendix 1: section 6, written by Fengjun Jin and Chengjin Wang), ecological environment and resource system (Appendix 1: section 7, written by Suocheng Dong and Yu Li), and regional management system and mechanism (Appendix 1: section 8, written by Xiaolu Gao and Wei Sun).

The third part is regional study. By making full use of the advantages of Nanjing Institute of Geography and Limnology, Northeast Institute of Geography and Agroecology, Chengdu Institute of Mountain Hazards and Environment and Xinjiang Institute of Ecology and Geography of CAS in the research on different types of regional development issues, the roadmap design for the regional development of the developed eastern coastal areas (Appendix 2: section1, written by Guishan Yang and Wen Chen), old industrial bases (Appendix 2:section2, written by Pingyu Zhang and Yanji Ma), mountainous area (Appendix 2:section3, written by Wei Deng and Yiping Fang), and border areas (Appendix 2:section4, written by Xiaolei Zhang and Zhaoping Yang) is worked out.

In the meantime, regional development experts outside of the CAS are also retained and involved in the consultation and discussion. A research report of approximately 500,000 words (which will also be published as an academic monograph) is completed, based on which this report takes shape.

This project embodies the painstaking efforts of many leaders and experts. CAS President Yongxiang Lu , together with other leaders of CAS, has heard reports twice and given specific instructions on the launching of this project. Director Jiaofeng Pan, section chief Feng Zhang, and Wenyuan Wang and Cheng Tao from the CAS Bureau of Planning and Strategy have made meticulous arrangements and provided valuable assistances. Director Bojie Fu and vice-director Renguo Feng from Resources and Environment Bureau have attended our discussions many times, proposing helpful suggestions from the overall perspective of resources and environment. Besides, experts in and outside CAS, such as Zhixiang She, Xiaogan Yu, Kongjian Yu, Shangyi Zhou, and Zheng Wang are also engaged in and make contributions to our project. Also, director Yi Liu and vice-director Guirui Yu of IGSNRR (Institute of Geographic Sciences and Natural Resources Research, CAS) and Prof. Hanying Mao have done much specific work for this project. Hereby, we would like to extend our sincerest thanks to all of them!

When this project started, Academician Chuanjun Wu attended the seminar in person and gave instructions. However, before the fruits of this project being published, he has left us forever. Nevertheless, the interdisciplinary nature of geography pointed out by him is still goes on, the discipline construction orientation targeting national demands proposed by him still goes on, and the cause of bringing Chinese issues into the world led by him still goes on. We deeply cherish the memory of Academician Wu!

It is worth mentioning that the report puts forward the establishment of the "Simulation and Decision Support System for Regional Sustainable Development in China" at Chinese Academy of Sciences. At present, under the care of CAS President Yongxiang Lu and with the specific instructions of vice-president Zhongli Ding, the pre-research project of the system construction is soon to be launched. Along with the implementation of the Roadmap of Regional Development, the CAS will consistently enhance its technology supporting capability in the research field of regional development to better serve the society, and the disciplinary construction of regional development research will also reach international advanced levels.